Artificial Intelligence for Natural Language Processing

Artificial Intelligence for Natural Language Processing offers a comprehensive exploration of how advanced computational methods are transforming the way machines understand human language. This book delves into the core principles of Natural Language Processing through an engaging progression – from fundamental word-level analysis to complex discourse and pragmatic analysis – integrating linguistic theory with cutting-edge Artificial Intelligence methodologies. It provides a robust framework for both the theoretical underpinnings and practical applications of NLP, ensuring that readers gain a clear understanding of how computers can effectively process and interpret human language.

What sets this book apart is its methodical structure that guides the reader through each level of language analysis, building upon earlier chapters to culminate in a deep integration of artificial intelligence within NLP systems. The detailed explanations and examples are designed to bridge the gap between abstract theory and real-world application, making it an invaluable resource for anyone looking to grasp the nuances of language processing.

FEATURES

- Provides a step-by-step progression from word-level analysis to syntactic, semantic, and pragmatic processing
- Offers in-depth discussions on word sense disambiguation with illustrative examples
- Presents an exploration of discourse integration and contextual meaning essential for modern NLP models
- Delivers comprehensive coverage of AI applications in NLP, highlighting state-of-the-art computational techniques
- Suggests clear, accessible explanations suitable for both beginners and advanced practitioners

This book is ideal for graduate students, researchers, and professionals in computer science, linguistics, and artificial intelligence. Whether you are a seasoned researcher looking to deepen your understanding or a newcomer eager to explore the field, *Artificial Intelligence for Natural Language Processing* serves as both an essential academic resource and a practical guide for navigating the evolving landscape of language technology.

Artificial Intelligence for Natural Language Processing

Dhanalekshmi Prasad Yedurkar,
Ganesh R. Pathak, Manisha Galphade,
and Thompson Stephan

CRC Press
Taylor & Francis Group
Boca Raton London New York

CRC Press is an imprint of the
Taylor & Francis Group, an **informa** business

Designed cover image: Shutterstock Image ID 2531574919

First edition published 2026
by CRC Press
2385 NW Executive Center Drive, Suite 320, Boca Raton FL 33431

and by CRC Press
4 Park Square, Milton Park, Abingdon, Oxon, OX14 4RN

CRC Press is an imprint of Taylor & Francis Group, LLC

ISBN: 978-1-032-54530-1 (hbk)
ISBN: 978-1-032-54532-5 (pbk)
ISBN: 978-1-003-42532-8 (ebk)

DOI: 10.1201/9781003425328

Typeset in Caslon
by codeMantra

Contents

Preface

When we first began exploring the world of Natural Language Processing, we were struck by how effortlessly human beings communicate – sharing nuanced meanings, emotions, and ideas through language. This book is a product of our longstanding fascination with the intricate processes behind human communication and the challenge of replicating that understanding in machines. We wanted to create a resource that not only breaks down the technical components of NLP but also paints a broader picture of how these components interact within a framework driven by artificial intelligence.

At its core, this book is about connecting the dots between language, meaning, and technology. We have always believed that the key to advancing NLP lies in understanding both its linguistic roots and its computational applications. Throughout the chapters, we discuss everything from basic word-level analysis to more advanced topics like syntactic parsing, semantic disambiguation, and pragmatic analysis. Our aim is to guide you through the progressive layers of NLP, revealing how each level builds upon the previous one and how artificial intelligence ties it all together.

Our motivation for writing this book was twofold. First, we wanted to demystify the subject for students, researchers, and professionals who are keen to dive deeper into the field but may find existing literature either too theoretical or overly technical. Second, we wanted to share insights gleaned from both academic research and practical applications, offering a balanced perspective that encourages readers to see the relevance of NLP in solving real-world problems. In writing this book, we hope to inspire a deeper appreciation for the complexities of language and the innovative methods used to decode it.

We invite you to explore these pages with curiosity and an open mind, as we journey together into the heart of language processing and artificial intelligence. Whether you are embarking on your first exploration of NLP or seeking to refine your understanding of its advanced applications, this book is meant to be a practical, engaging, and thought-provoking guide.

Authors

Dhanalekshmi Prasad Yedurkar is a Postdoctoral Researcher at the University of Augsburg, Germany. She is also affiliated to MIT Art, Design and Technology University, Pune, India, as an Associate Professor in the School of Computing. She earned a PhD on the application of digital signal processing and artificial intelligence in anomaly detection of EEG signal. Her current research interests include natural language processing, tool condition monitoring for CNC processes, diagnosis and prognosis of gear monitoring, biomedical signal processing, machine learning, Internet of Things, and image processing.

Ganesh R. Pathak is an Academic and a Researcher with over 26 years of experience bridging academia and industry. He is currently a Professor and Head of the Department of Computer Science and Engineering at the School of Computing, MIT Art, Design and Technology University, Pune.

He earned a PhD in computer science and engineering with his research focused on developing a security framework for wireless sensor networks. He has published 34 research papers in reputed peer-reviewed journals and conference proceedings, many of which are indexed in Scopus and SCI. His research and teaching interests focus on artificial intelligence, big data analytics, and cognitive modelling. As a mentor, he guides doctoral candidates in artificial intelligence, data science, cloud computing, and security.

Beyond his research, he is an active contributor to the academic and professional community. He serves on various boards of studies, and he contributes as a reviewer and session chair at national and international conferences. Dr. Pathak also promotes automation and e-governance in educational processes within the university. His efforts extend to skill development through numerous organized workshops, seminars, and faculty development programs, which strengthen his role as an admonitor in academia.

Through his scholarly achievements, project leadership, and engagement in the academic community, Dr. Pathak continues to make contributions to the university and various institutions in advancing technology education and research.

Manisha Galphade is an experienced Academic Professional currently working at the School of Computing, MIT Art, Design and Technology University, Pune. With a teaching career spanning 16 years, she has contributed significantly to the field of education. She teaches subjects including machine learning, database management system, theory of computation, data mining, and many more. Throughout her career, she has authored four conference papers, five journal papers, and three book chapters, showcasing her dedication to research and academic development. She is also pursuing a PhD at Veermata Jijabai Technological Institute, Mumbai, furthering her academic expertise. Her research area is artificial intelligence with research interests in time series analysis, image processing, and signal processing. Passionate about fostering innovation and critical thinking, she strives to inspire students to reach their full potential. Her extensive experience and research contributions reflect her commitment to advancing knowledge and fostering growth in her field.

Thompson Stephan earned a PhD at Pondicherry University, India, in 2018, and he has nearly 7 years of academic experience, complemented by full-time research and industry expertise. He serves as an Assistant Professor at the Thumbay College of Management and AI in Healthcare, Gulf Medical University, Ajman, United Arab Emirates. Recognized among Stanford/Elsevier's Top 2% Scientists globally in both 2023 and 2024, Dr. Thompson has received prestigious accolades, including the Best Researcher Award in 2020 and the Protsahan Research Award in 2023, both from the IEEE Bangalore Section, India. His primary research focus is artificial intelligence with specialized expertise in advancing machine learning, data mining, and meta-heuristic optimization. With more than 80 Scopus-indexed publications, including 48 in SCI-indexed journals, Thompson's work has garnered significant recognition. He actively contributes as a book editor and reviewer for esteemed international journals, with publications on leading platforms such as IEEE, Elsevier, Taylor & Francis, and Springer.

1

INTRODUCTION AND WORD-LEVEL ANALYSIS

Natural Language Processing (NLP) is a domain within Computer Science and Engineering, particularly in Artificial Intelligence (AI), focused on a computer's capacity to comprehend, interpret, analyze, and alter any human language. It emphasizes the communication between computers and humans through natural language. It is a part of AI dedicated to interpreting human language in both written and spoken forms. NLP allows computers to comprehend, interpret, and produce human language that is contextually significant and meaningful.

1.1 History of NLP

- **Machine Translation Phase (1940–1960):** Early attempts at automated language translation, with rule-based systems and limited linguistic understanding [1].
- **AI Influenced Phase (1960s–1970):** Shift towards AI, incorporating symbolic reasoning and knowledge representation into language processing.
- **Grammatico-logical Phase (1970–late 1980):** Development of rule-based systems that focused on syntactic and semantic parsing for more in-depth language understanding.
- **Lexical & Corpus Phase (1990):** Greater reliance on statistical methods, corpus linguistics, and lexical resources for language processing.

DOI: 10.1201/9781003425328-1

1.2 Generic NLP System

Generic NLP system includes various components and processes:

- **Input**: The system takes in input data in the form of text, speech, or a combination of both.
- **Natural Language Understanding (NLU)**: The NLU component analyzes the input to extract meaning, intent, and context.

  Tasks include parsing sentences, identifying entities (such as people, places, and things), determining sentiment, and understanding the relationships between words and phrases.
- **Processing and Inference**: The system processes the extracted information, often using machine learning algorithms or rule-based systems, to make inferences and derive insights.
- **Natural Language Generation (NLG)**: NLG is the process of producing meaningful and coherent language output. It involves generating human-like text or speech based on the understanding derived from NLU or other sources. This encompasses tasks such as creating summaries, generating responses, language translation, and constructing coherent paragraphs or articles.
- **Output**: The final output is generated text, speech, or actionable information, depending on the application

1.3 Ambiguity and Challenges

Ambiguity is the primary challenge in NLP, as words possess distinct meanings that vary according to context, resulting in ambiguity at lexical, syntactic, and semantic levels. Natural language is inherently ambiguous, as it can be understood in multiple ways [2].

NLP has various types of ambiguities:

- **Lexical**: Ambiguity due to multiple meanings of a single word.
- **Syntactic**: Ambiguity arising from multiple possible interpretations of the grammatical structure of a sentence.
- **Semantic**: Ambiguity from words or phrases having multiple meanings based on context.

- **Pragmatic**: Ambiguity arising from multiple interpretations of a context of a phrase.
- **Anaphoric**: Ambiguity occurring when a pronoun or expression refers to more than one possible antecedent.
- **Structural Ambiguity**: Ambiguity in the overall structure or organization of a sentence, leading to different interpretations.

1.4 Words

In NLP, the concept of "words" plays a central role, but it encompasses various meanings depending on the context. At its core, a word is often treated as a token, which is the smallest unit of text that carries meaning. For instance, in the sentence "NLP is fascinating!," the words are tokenized into ["NLP," "is," "fascinating," "!"]. Words can be classified into different types based on their role in text. Content words such as nouns, verbs, and adjectives carry semantic meaning, while function words like prepositions and conjunctions primarily serve grammatical purposes. Preprocessing often involves removing stopwords, which are common but less meaningful words like "and," "the," or "is."

Words in NLP are not just textual entities but are also represented mathematically. Word embeddings, including GloVe and Word2Vec, are numerical vector representations that include semantic and syntactic links among words. Additionally, techniques like Byte Pair Encoding (BPE) or SentencePiece break words into smaller units or subword tokens, enabling models to handle complex words and inflected forms effectively. However, working with words presents challenges such as ambiguity (e.g., the word "bank" can mean side of a river or a financial institution), handling out-of-vocabulary (OOV) words not seen during training, and managing morphological variations like "run," "running," or "runs."

NLP tasks such as part-of-speech tagging assign grammatical categories to words, while named entity recognition identifies specific entities like names, dates, or locations. Advanced techniques such as word sense disambiguation aim to determine the correct meaning of a word based on its context. Despite being fundamental, words pose numerous challenges, especially in languages with complex structures, requiring sophisticated methods to ensure accurate representation and understanding in NLP applications.

1.5 Corpora

In NLP, a **corpus** (plural: **corpora**) refers to a large and structured collection of texts that serve as a foundation for linguistic analysis and model training. These corpora are essential resources, providing raw material for various NLP tasks, from language modeling to machine translation and sentiment analysis. A corpus can be as simple as a collection of plain text documents or as complex as annotated datasets enriched with additional metadata, such as POS tags, syntactic structures, or semantic roles.

Corpora are often categorized based on their content and purpose. **General-purpose corpora**, such as the British National Corpus (BNC) or Wikipedia, contain diverse and broad text types, making them suitable for generic language tasks. **Domain-specific corpora**, on the other hand, focus on specialized fields like legal, medical, or scientific texts, supporting tasks that require domain expertise. Another classification is based on annotations. **Raw corpora** are unprocessed text collections, while **annotated corpora** include labels or tags for tasks such as sentiment analysis, NER, or syntactic parsing. Popular annotated corpora include the Penn Treebank for syntactic parsing and the CoNLL datasets for NER.

Some corpora are multilingual, supporting tasks like machine translation or cross-lingual language modeling. Examples include the Europarl corpus, which contains parliamentary proceedings in multiple European languages, and the UN Parallel Corpus, which offers translated United Nations documents. Additionally, corpora can be dynamic and ever-expanding, such as social media datasets like the Twitter corpus, reflecting real-time trends and conversational language. However, building and using corpora come with challenges, including data privacy, ethical concerns, and ensuring representativeness across languages and demographics. Despite these challenges, corpora remain indispensable in NLP, driving advancements in language understanding, generation, and analysis.

1.6 Phases of NLP

NLP is divided into five principal stages [3], commencing with basic word processing and advancing to the interpretation of complex phrase meanings, as illustrated in Figure 1.1.

1. Lexical/Morphological Analysis
2. Syntax Analysis or Parsing
3. Semantic Analysis
4. Pragmatic Analysis
5. Discourse Integration

1.6.1 Morphological/Lexical Analysis

This is the preliminary phase in NLP, concentrating on identifying and analyzing word structures. The aggregate of words and phrases in a language is termed the lexicon. Lexical analysis is the procedure of decomposing a text file into paragraphs, phrases, and words, transforming source code into comprehensible lexemes. This process searches for morphemes, the smallest units of a word, and identifies their relationships, transforming the word into its root form and assigning probable parts of speech (POS). Two basic types of morphological composition are as follows.

- **Inflectional Morphology:** It generates many variants of the same word, conveying certain grammatical information without altering the fundamental meaning. It adds information to a word consistent with its context within a sentence

    E.g. automaton → automata

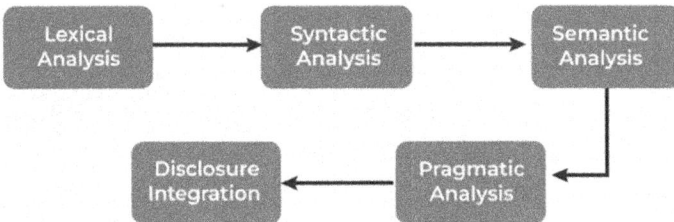

Figure 1.1   Phases of NLP.

- **Derivational Morphology**: It generates various words from the same root, including distinct meanings, grammatical categories, and frequently introducing new parts of speech.

   E.g. parse → parser

1.6.2 Syntax Analysis or Parsing

The **syntax analysis** or **parsing phase** in NLP focuses on analyzing the grammatical structure of a given sentence to determine how words are arranged and related according to the rules of a language. This phase ensures that the input text adheres to the syntax of the language being processed. The primary goal of parsing is to construct a **parse tree** or **syntax tree**, which represents the hierarchical structure of the sentence, showing the relationships between words and phrases. For instance, in the sentence *"The cat sat on the mat,"* parsing identifies components such as the noun phrase (*"The cat"*) and the verb phrase (*"sat on the mat"*) and their roles in the sentence.

Parsing can be performed using various techniques, such as **dependency parsing**, which focuses on relationships between words, and **constituency parsing**, which breaks sentences into nested sub-phrases based on grammatical rules. Syntax analysis is crucial for higher-level NLP tasks like semantic analysis, where meaning is derived from syntactic structures. It also helps in identifying errors in the text, such as grammatical mistakes, and is a foundational step for applications like machine translation, question answering, and text generation. Despite its importance, syntax analysis faces challenges, especially with ambiguous sentences where multiple valid parse trees can exist, requiring sophisticated algorithms to resolve such ambiguities accurately.

1.6.3 Semantic Analysis

Semantic analysis in NLP is the phase where the focus shifts from the structure of a sentence to its meaning. It aims to understand the literal meaning of words, phrases, and sentences within a given context, ensuring that the text's interpretation aligns with its intended sense. This phase involves tasks such as resolving word meanings (lexical semantics), analyzing the relationships between words (semantic roles), and constructing a logical representation of the text. For example, in the sentence *"John gave Mary a book,"* semantic analysis identifies the entities (*John*, *Mary*, and *book*), their roles (giver, receiver, and object), and the action (*gave*).

A critical aspect of semantic analysis is **word sense disambiguation**, which resolves ambiguities by determining the correct meaning of a word based on its context. Another important task is identifying **semantic relationships**, such as synonyms, antonyms, and hierarchical relations (e.g., a *dog* is a type of *animal*). Semantic analysis frequently employs methodologies such as semantic networks, ontologies, or pre-trained language models to construct a coherent representation of the text. It plays a vital role in applications such as question-answering systems, machine translation, and information retrieval, where understanding meaning is crucial. However, challenges arise from figurative language, idioms, and context-dependent nuances, making semantic analysis one of the most complex and significant phases of NLP.

1.6.4 Discourse Integration

Discourse integration is a critical phase in NLP that goes beyond understanding individual sentences to analyze the relationships and coherence between sentences in a larger context, such as paragraphs or entire documents. It ensures that the meaning of a sentence is interpreted in light of the surrounding text, recognizing that sentences do not exist in isolation but are part of a broader discourse. For example, in the two sentences *"John bought a car. He loves it,"* discourse integration identifies that *"he"* refers to John and *"it"* refers to the car, establishing semantic connections across sentences.

This phase involves tasks like **coreference resolution**, which identifies entities that refer to the same object or person across sentences, and **anaphora resolution**, which connects pronouns or other referring expressions to their antecedents. Discourse integration also examines how ideas flow logically, identifying relationships such as causality, contrast, or elaboration. For instance, in *"She studied hard because she wanted to pass,"* the causal relationship between studying and wanting to pass is established.

Discourse integration is crucial for tasks like summarization, question answering, and dialogue systems, where understanding the overall context and continuity of ideas is essential. However, this phase can be challenging due to ambiguities, implied information, and the variability of natural language, requiring sophisticated models to effectively capture and utilize contextual information.

1.6.5 Pragmatic Analysis

Pragmatic analysis is the final phase of NLP that focuses on interpreting the intended meaning of text or speech by considering the context, speaker's intentions, and situational factors. Unlike earlier phases that deal with literal meanings or syntactic structures, pragmatic analysis examines the **implied meaning** behind words and phrases, understanding nuances like sarcasm, humor, politeness, or indirect requests. For instance, in the sentence *"Can you pass the salt?"* the literal meaning is a question about ability, but pragmatically, it is interpreted as a polite request to pass the salt.

This phase considers aspects such as the **speaker's intent**, the relationship between participants in a conversation, cultural norms, and the environment in which communication takes place. Pragmatic analysis often involves tasks like **speech act recognition**, where the type of communication (e.g., request, command, or question) is identified, and **deixis resolution**, which interprets words like *"this," "that," "here,"* and *"there"* based on context.

Pragmatic analysis is essential for applications like conversational AI, sentiment analysis, and machine translation, where understanding context-dependent meaning is critical for accurate and human-like responses. However, this phase is challenging due to the complexities of human communication, such as ambiguous expressions, cultural differences, and the need to infer unstated intentions or shared knowledge, making it one of the most sophisticated tasks in NLP.

1.7 Basic Concepts of Text Preprocessing

Text preprocessing [4] is an essential phase in NLP that prepares unrefined text for examination and modeling. It involves transforming unstructured text into a structured format suitable for computational processing. Techniques such as **tokenization, stemming, lemmatization, normalization, and Bag of Words** (BoW) are used to clean and simplify text, ensuring consistency [5]. Additionally, advanced tools like **regular expressions, finite automata, finite-state transducers**, and **n-gram language models** help extract patterns and understand context. Effective preprocessing improves the performance of NLP models by reducing noise, standardizing data, and capturing meaningful features from text.

1.7.1 Stemming

Stemming is transforming words into their base forms according to specific rules, regardless of significance. It is reducing words to their base or root form. The most common algorithm for stemming is Porters Algorithm

For example – troubled, troubles → trouble

article → articl

Implementation –

```
% Python program

from nltk.stem import PorterStemmer

def stemming_example(word):
    ps = PorterStemmer()
    stemmed_word = ps.stem(word)
    return stemmed_word

word1 = "running"
word2 = "connected"
word3 = "troubled"

result1 = stemming_example(word1)
result2 = stemming_example(word2)
result3 = stemming_example(word3)

print('Stemmed word of', word1, ':', result1)
print('Stemmed word of', word2, ':', result2)
print('Stemmed word of', word3, ':', result3)
```

% Output:

Stemmed word of running : run
Stemmed word of connected : connect
Stemmed word of troubled : troubl

% Python program

```
from nltk.stem import PorterStemmer
import pandas as pd

# Initialize the stemmer
porter_stemmer = PorterStemmer() # Corrected the typo
in initialization

# Words to stem
words = ["troubled", "running", "article",
"connecting", "lives"]

# Stem the words
stemmed_words = [porter_stemmer.stem(word) for word
in words]

# Create a DataFrame to show original and stemmed
words
stemdf = pd.DataFrame({'original_word': words,
'stemmed_word': stemmed_words})

# Display the DataFrame
print(stemdf)
```

% Output:

	original_word	stemmed_word
0	troubled	troubl
1	running	run
2	article	articl
3	connecting	connect
4	lives	live

1.7.2 Lemmatization

Lemmatization is transforming words into their base forms through vocabulary mapping. It is similar to stemming but considers the word's meaning, transforms words to actual the root. It uses parts of speech and its meaning.

But lemmatization is slower than stemming.

For example – better → good

Implementation

```
% Python program

from nltk.stem import WordNetLemmatizer

def lemmatization_example(word, pos='n'):
    lemmatizer = WordNetLemmatizer()  # Initialize
lemmatizer
    lemmatized_word = lemmatizer.lemmatize(word,
pos=pos)  # Corrected variable name
    return lemmatized_word  # Return the lemmatized word

# Example words
word1 = "running"
word2 = "article"
word3 = "troubled"

# Get the lemmatized words
result1 = lemmatization_example(word1)  # Corrected
variable assignment
result2 = lemmatization_example(word2)  # Fixed the
result variable
result3 = lemmatization_example(word3)  # Corrected
the result variable

# Print results
print('Lemmatized word of', word1, ':', result1)  #
Corrected print statement syntax
print('Lemmatized word of', word2, ':', result2)
print('Lemmatized word of', word3, ':', result3)
```

% Output:

```
Lemmatized word of running : running
Lemmatized word of article : article
Lemmatized word of troubled : troubled
```

% Python program

```python
from nltk.stem import WordNetLemmatizer
import pandas as pd

# Initialize lemmatizer
lemmatizer = WordNetLemmatizer()

# Words to lemmatize
words = ["troubled", "running", "article",
"connecting"]

# Lemmatize the words (using part-of-speech 'v' for
verbs)
lemmatized_words = [lemmatizer.lemmatize(word,
pos='v') for word in words]

# Create DataFrame to show original and lemmatized
words
lemmatizeddf = pd.DataFrame({'original_word': words,
'lemmatized_word': lemmatized_words})

# Display the DataFrame
print(lemmatizeddf)
```

% Output:

	original_word	stemmed_word
0	troubled	troubl
1	running	run
2	article	articl
3	connecting	connect

1.7.3 Normalization

Text normalization is transformation of text into a standard (canonical) form. It ensures consistency in text data. Text normalization is beneficial for noisy texts, including blog article comments, text messages, and social media comments, where Out-Of-Vocabulary (OOV) phrases, misspellings, and abbreviations are frequent.

```python
% Python program

import re
import pandas as pd

def normalize_text(text):
    # Convert to lowercase
    text = text.lower()

    # Replace repeated characters (e.g., gooood to good)
    text = re.sub(r'(.)\1+', r'\1', text)

    # Replace common abbreviations
    text = re.sub(r'\bgud\b', 'good', text)
    text = re.sub(r'\bgr8\b', 'great', text)

    # Remove special characters, numbers, and
punctuations
    text = re.sub(r'[^a-zA-Z\s]', '', text)

    # Remove extra whitespaces
    text = re.sub(r'\s+', ' ', text).strip()

    return text

# Example texts
texts = ["stop-words", "gud", "This is gr8!"]

# Normalize the texts
normalized_texts = [normalize_text(text) for text in
texts]
```

```
# Create DataFrame
df = pd.DataFrame({'Original Text': texts, 'Normalized
Text': normalized_texts})

# Display the DataFrame
print(df)

% Output:
```

	Original Text	Normalized Text
0	stop-words	stopwords
1	gud	good
2	This is gr8!	this is great

1.7.4 Tokenization

Tokenization refers to the process of dividing text into meaningful units, specifically words. Both phrase tokenizers and word tokenizers exist. The sentence tokenizer segments a paragraph into coherent sentences, whereas the word tokenizer partitions the phrase into individual meaningful words.

Implementation –

```
% Python program

from nltk.tokenize import word_tokenize, sent_tokenize

def tokenization_example(text):
    # Tokenize words
    words = word_tokenize(text)

    # Tokenize sentences
    sentences = sent_tokenize(text)

    return words, sentences

# Example text to tokenize
text_to_tokenize = "This is a sample sentence.
Tokenize it."
```

```
# Get word and sentence tokens
word_tokens, sentence_tokens =
tokenization_example(text_to_tokenize)

# Print results
print(f"Word tokens: {word_tokens}")
print(f"Sentence tokens: {sentence_tokens}")

% Output:

Word tokens: ['This', 'is', 'a', 'sample', 'sentence',
'.', 'Tokenize', 'it', '.']
Sentence tokens: ['This is a sample sentence.',
'Tokenize it.']
```

1.7.5 *Bag of Words*

The BoW is a method employed in NLP to represent textual data as a collection of numerical attributes. In this framework, each document or text segment is presented as a "bag" of words, with each word denoted by a distinct feature or dimension in the resultant vector. The value of each attribute is calculated by the frequency of the relevant word's occurrence in the text. The BoW technique is utilized for feature extraction from textual materials.

Implementation –

```
% Python program

import pandas as pd
import numpy as np
import re

# Sample documents
doc1 = "Natural Language Processing is fascinating!"
doc2 = "NLP applications are widespread."
doc3 = "Text analysis and machine learning are
essential in NLP."

# Tokenize and normalize the words in each document
doc1_words = re.sub(r"[^a-zA-Z0-9]", " ", doc1.
lower()).split()
doc2_words = re.sub(r"[^a-zA-Z0-9]", " ", doc2.
lower()).split()
```

```
doc3_words = re.sub(r"[^a-zA-Z0-9]", " ", doc3.
lower()).split()

# Create a set of unique words across all documents
wordset12 = np.union1d(doc1_words, doc2_words)
wordset = np.union1d(wordset12, doc3_words)

# Function to calculate Bag of Words representation
for a document
def calculateBOW(wordset, doc_words):
    tf_dict = dict.fromkeys(wordset, 0)  # Initialize
dictionary with 0 for all words in wordset
    for word in doc_words:
        tf_dict[word] += 1  # Increment count for each
word in the document
    return tf_dict

# Calculate Bag of Words for each document
bow1 = calculateBOW(wordset, doc1_words)
bow2 = calculateBOW(wordset, doc2_words)
bow3 = calculateBOW(wordset, doc3_words)

# Create a DataFrame to represent the Bag of Words for
each document
df_bow = pd.DataFrame([bow1, bow2, bow3])

# Display the DataFrame
print(df_bow.head())

% Output:
```

	analysis	and	applications	are	essential	fascinating	in	is	language	\
0	0	0	0	0	0	1	0	1	1
1	0	0	1	1	0	0	0	0	0
2	1	1	0	1	1	0	1	0	0

	learning	machine	natural	nlp	processing	text	widespread
0	0	0	1	0	1	0	0
1	0	0	0	1	0	0	
2	1	1	0	1	0	1	0

1.7.6 Regular Expression

Regular Expression (RE) [6], a language for specifying text strings, consists of basic units such as characters or strings. It is a standard notation for representation of text sequences.

RE helps to find or match other strings or set of strings, to define a pattern to search through a corpus.

Regular expressions are implemented by finite-state automaton.

An expression written using the set of operators (+, ., *) and describing a regular language is known as regular expression.

Example : $(0 + 10^*) - \{0, 1, 10, 100, 1000, 10000, \dots \}$

Implementation

```
% Python program

import re

# Example 1: Matching a pattern in a string
text = "Natural Language Processing is a key aspect of
modern AI."
pattern = r'\b\w{6}\b'  # Matches six-letter words

matches = re.findall(pattern, text)  # Corrected the
assignment
print(f"Matches: {matches}")

# Example 2: Replacing a pattern in a string
text = "Hello, world! This is an example sentence."
pattern = r'\b\w{5}\b'  # Matches five-letter words
replacement = "***"

modified_text = re.sub(pattern, replacement, text)
print(f"\nModified Text: {modified_text}")

% Output:

Matches: ['aspect', 'modern']

Modified Text: ***, ***! This is an example sentence.

% Python program

import re
```

```python
# Example 1: Using re.match()
match_result = re.match(r'^Hello', 'Hello, World!')
print(f"Match: {match_result.group() if match_result
else 'No match'}")

# Example 2: Using re.search()
search_result = re.search(r'World', 'Hello, World!')
print(f"Search: {search_result.group() if search_
result else 'Not found'}")

# Example 3: Using re.findall()
words = re.findall(r'\b\w+\b', 'Hello, World!')
print(f"Find all words: {words}")

# Example 4: Using re.sub()
modified_text = re.sub(r'World', 'Universe', 'Hello,
World!')
print(f"Replace: {modified_text}")

# Example 5: Using re.split()
tokens = re.split(r'\s', 'This is a sample sentence.')
print(f"Split: {tokens}")

% Output:

Match: Hello
Search: World
Find all words: ['Hello', 'World']
Replace: Hello, Universe!
Split: ['This', 'is', 'a', 'sample', 'sentence.']
```

1.7.7 Finite-State Automaton (FSA)

An automaton is an abstract model of a computer. Finite-State Automaton (FSA) [7] is a significant tool of computational linguistics. An automaton having a finite number of states is called a FSA or Finite Automaton (FA). The set of languages that can be characterized by FSAs are called regular expressions.

Mathematically, an automaton can be represented by a 5-tuple $(Q, \Sigma, \delta, q0, F)$, where –

- Σ is a finite set of symbols, called the alphabet of the automaton.
- q0 is the initial state from where any input is processed ($q0 \in Q$).
- Q is a finite set of states.
- δ is the transition function
- F is a set of final state/states of Q ($F \subseteq Q$).

Implementation

```
% Python program

class FiniteAutomaton:
    def __init__(self, states, alphabet, transitions,
start_state, accepting_states):
        self.states = states
        self.alphabet = alphabet
        self.transitions = transitions
        self.current_state = start_state
        self.accepting_states = accepting_states

    def process_input(self, input_string):
        for symbol in input_string:
            if symbol not in self.alphabet:
                return False  # Reject if symbol is
not in the alphabet
            # Transition to next state based on
current state and input symbol
            self.current_state = self.transitions.
get((self.current_state, symbol), None)
            if self.current_state is None:
                return False  # Reject if no valid
transition

        # Accept if the final state is an accepting
state
        return self.current_state in self.
accepting_states

# Example: A simple DFA to recognize 'ab'
states = {'q0', 'q1'}
alphabet = {'a', 'b'}
transitions = {('q0', 'a'): 'q1', ('q1', 'b'): 'q0'}
```

```
start_state = 'q0'
accepting_states = {'q0'}

dfa = FiniteAutomaton(states, alphabet, transitions,
start_state, accepting_states)

input_str = 'ababab'
result = dfa.process_input(input_str)

print(f"Does the input '{input_str}' match the
pattern? - {result}")

% Output:

Does the input 'ababab' match the pattern? - True
```

1.7.8 Finite-State Transducer

A FSA represents a set of strings, a regular language.

E.g. {walk, walks, walked, love loves, loved}

A Finite-State Transducer (FST) represents a set of pairs of strings (as input and output pairs)

{(walk, walk+V+PL), (walk, walk+N+SG), (walked, walk+V+PAST)...}

FSA have input labels – one input tape

FST have input:output pairs on labels – two tapes: input and output.

Mathematically, a finite-state transducer $T = \langle Q, \Sigma, \Delta, q0, F, \delta, \sigma \rangle$ consists of:

- A finite alphabet Σ of input symbols (e.g. $\Sigma = \{a, b, c,...\}$)
- A designated start state $q0 \in Q$
- A finite set of states $Q = \{q0, q1,.., qn\}$
- A finite alphabet Δ of output symbols (e.g. $\Delta = \{+N, +pl,...\}$)
- A set of final states $F \subseteq Q$
- A transition function δ: $Q \times \Sigma \rightarrow 2Q$ [$\delta(q,w) = Q'$ for $q \in Q$, $Q' \subseteq Q$, $w \in \Sigma$]
- An output function σ: $Q \times \Sigma \rightarrow \Delta^*$ [$\sigma(q,w) = \omega$ for $q \in Q$, $w \in \Sigma$, $\omega \in \Delta^*$]

Implementation

```
% Python program

class FiniteStateTransducer:
    def __init__(self, states, alphabet, transitions,
start_state, final_states):
        self.states = states
        self.alphabet = alphabet
        self.transitions = transitions
        self.current_state = start_state
        self.final_states = final_states

    def process_input(self, input_string):
        for symbol in input_string:
            if symbol not in self.alphabet:
                print(f"Invalid symbol: {symbol}")
                return False

            transition_key = (self.current_state,
symbol)
            if transition_key in self.transitions:
                self.current_state = self.
transitions[transition_key]
            else:
                print(f"No transition defined for
{transition_key}")
                return False

        return self.current_state in self.final_states


# Example: A simple FST to recognize 'ab' or 'cd'
states = {'q0', 'q1'}
alphabet = {'a', 'b', 'c', 'd'}
transitions = {('q0', 'a'): 'q1', ('q1', 'b'): 'q0',
('q0', 'c'): 'q1', ('q1', 'd'): 'q0'}
start_state = 'q0'
final_states = {'q0'}

fst = FiniteStateTransducer(states, alphabet,
transitions, start_state, final_states)
```

```
# Test the FST
input_str1 = 'abcd'
input_str2 = 'acbd'

result1 = fst.process_input(input_str1)
result2 = fst.process_input(input_str2)

# Display the results
print(f"Does '{input_str1}' match the pattern?
{result1}")
print(f"Does '{input_str2}' match the pattern?
{result2}")

% Output:

No transition defined for ('q1', 'c')
Does 'abcd' match the pattern? True
Does 'acbd' match the pattern? False
```

1.7.9 N-Gram Language Model

Language modeling is the way of determining the probability of any sequence of words. Language models calculate the probabilities of a sentence or series of words, as well as the likelihood of a subsequent word based on a preceding set of words. N-grams [8] are consecutive sequences of elements derived from a text or audio corpus, or virtually any form of data. The variable n in n-grams denotes the quantity of elements to evaluate: unigram for $n=1$, bigram for $n=2$, trigram for $n=3$, and so forth. n-gram and n-gram models are widely used in probability, communication theory, computational linguistics like statistical NLP, computational biology etc.

1.7.9.1 Steps in N-Grams Language Model

1. **Data Preparation:** Tokenize the text into words or other meaningful units.
2. **Choose N:** Determine the value of N for the desired context (e.g., unigrams, bigrams, trigrams).

3. **Generate N-Grams:** Create sequences of N consecutive words from the tokenized text.
4. **Count Occurrences:** Count the frequency of each n-gram in the dataset.
5. **Calculate Probabilities:** Calculate the probability of the next word given the context of the N-1 preceding words.
6. **Smoothing (Optional):** Apply smoothing techniques to handle unseen n-grams and prevent zero probabilities.
7. **Build Language Model:** Use the n-grams and associated probabilities to build a language model

Formula – Conditional probabilities – $p(B|A) = P(A,B)/P(A)$
   Chain Rule – $P(A,B) = P(A)P(B|A)$
   in General – $P(x1,x2,x3,\ldots,xn) = P(x1)P(x2\,|x1)P(x3\,|x1,x2)\ldots$
$P(xn\,|x1,\ldots,xn\text{-}1)$

Implementation

```
% Python program

from nltk.util import ngrams
from nltk.tokenize import word_tokenize

# Example sentence
sentence = "This is a simple example for generating
bigrams."

# Tokenize the sentence into word tokens
tokens = word_tokenize(sentence)

# Define the size of N-grams (in this case, bigrams)
n = 2

# Generate bigrams
bigram_list = list(ngrams(tokens, n))

# Display the result
print(f"Original sentence: {sentence}")
print(f"Bigrams: {bigram_list}")
```

% Output:

Original sentence: This is a simple example for
generating bigrams.
Bigrams: [('This', 'is'), ('is', 'a'), ('a',
'simple'), ('simple', 'example'), ('example', 'for'),
('for', 'generating'), ('generating', 'bigrams'),
('bigrams', '.')]

% Python program

```python
import pandas as pd
from nltk.util import ngrams
from nltk.tokenize import word_tokenize

def generate_ngrams(text, n):
    # Tokenize the text
    tokens = word_tokenize(text)
    # Generate the n-grams
    n_grams = list(ngrams(tokens, n))
    return n_grams

def create_ngrams_dataframe(text, n):
    # Generate n-grams from the text
    n_grams = generate_ngrams(text, n)
    # Create a DataFrame from the n-grams
    n_grams_df = pd.DataFrame(n_grams, columns=
[f'word_{i}' for i in range(1, n+1)])
    return n_grams_df

# Example text
example_text = "This is a simple example for
generating N-grams."

# Specify the desired N for N-grams (e.g., 2 for
bigrams, 3 for trigrams)
desired_n = 4

# Generate and display N-grams in DataFrame
ngrams_dataframe = create_ngrams_dataframe(example_
text, desired_n)
print(ngrams_dataframe)
```

`% Output:`

	word_1	word_2	word_3	word_4
0	This	is	a	simple
1	is	a	simple	example
2	a	simple	example	for
3	simple	example	for	generating
4	example	for	generating	N-grams
5	for	generating	N-grams	.

1.8 Summary

This chapter provides a foundational introduction to NLP, exploring its history, fundamental concepts, and various stages of language processing. It discusses the generic NLP system, breaking it down into different levels and stages, highlighting the challenges posed by ambiguity in natural language. This chapter also introduces key linguistic components such as words, corpora, and morphology analysis, covering both inflectional and derivational morphology, which play a crucial role in understanding word structures and variations. The discussion extends to essential preprocessing techniques like stemming, lemmatization, and normalization, which are used to standardize text for computational analysis. Tokenization, an essential step in NLP, is explored in detail, followed by the BoW model, a popular method for text representation. This chapter also delves into pattern-matching techniques, including regular expressions, finite automata, and FST, which help in text parsing and recognition. Finally, it introduces n-grams language models, which are widely used for probabilistic text predictions and linguistic pattern analysis. This chapter sets the stage for deeper discussions in NLP by establishing the fundamental principles of word-level text processing and representation.

References

1. P. Johri, S. K. Khatri, A. T. Al-Taani, M. Sabharwal, S. Suvanov, and A. Kumar, "Natural language processing: History, evolution, application, and future work," in *Proceedings of 3rd International Conference on Computing Informatics and Networks: ICCIN 2020*, Delhi, India, 2021, pp. 365–375.

2. D. Khurana, A. Koli, K. Khatter, and S. Singh, "Natural language processing: state of the art, current trends and challenges," *Multimed. Tools Appl.*, vol. 82, no. 3, pp. 3713–3744, 2023.

3. T. P. Nagarhalli, V. Vaze, and N. K. Rana, "Impact of machine learning in natural language processing: A review," in *2021 Third International Conference on Intelligent Communication Technologies and Virtual Mobile Networks (ICICV)*, Tirunelveli, India, 2021, pp. 1529–1534, doi: 10.1109/ICICV50876.2021.9388621.

4. A. I. Kadhim, "An evaluation of preprocessing techniques for text classification," *Int. J. Comput. Sci. Inf. Secur.*, vol. 16, no. 6, pp. 22–32, 2018.

5. A. Tabassum and R. R. Patil, "A survey on text pre-processing & feature extraction techniques in natural language processing," *Int. Res. J. Eng. Technol.*, vol. 7, no. 06, pp. 4864–4867, 2020.

6. G. Kaur, "Usage of regular expressions in NLP," *Int. J. Res. Eng. Technol.*, vol. 3, no. 01, p. 7, 2014.

7. A. Maletti, "Survey: Finite-state technology in natural language processing," *Theor. Comput. Sci.*, vol. 679, pp. 2–17, 2017.

8. P. Majumder, M. Mitra, and B. B. Chaudhuri, "N-gram: a language independent approach to IR and NLP," in *International Conference on Universal Knowledge and Language* (ICUKL2002), Goa, India, 2002.

2
SYNTACTIC ANALYSIS

Syntactic processing involves examining a sentence's grammatical structure to comprehend its meaning. This entails recognizing the various parts of speech in a sentence, including adverbs, adjectives, verbs, and, nouns, and their interrelations to convey the sentence's intended meaning. Syntactic analysis is also known as parsing. In syntactic analysis, the objective is to comprehend the functions of each word inside the sentence, the interrelations among the words, and to analyze the grammatical structure of sentences to ascertain their accurate meaning.

In syntactic analysis, target is to:

- Identify the functions of words within a sentence,
- Analyze the connection among terms,
- Analyze the grammatical composition of sentences.

2.1 Parts of Speech Tagging

Parts of Speech tagging (POS Tagging) [1] involves identifying the parts of speech of each token. It is the process of annotating a word in a document to indicate its matching part of speech, determined by its definition and context. POS tagging, also known as speech tagging or word category disambiguation, helps understand the relationship between words, developing linguistic rules, and lemmatization.

Some common tags in nltk are

- CC – Coordinating conjunction
- CD – Cardinal Number
- DT – determiner
- FW – Foreign word

DOI: 10.1201/9781003425328-2

- IN – Preposition or Subordinating conjunction
- JJ – Adjective
- NN – Noun, singular
- NNS – Noun, plural
- NNP – Proper noun, singular
- NNPS – Proper noun, plural
- RP – Particle
- RB – Adverb
- VB – Verb, base form
- VBZ – Verb, 3rd person singular present
- VBP – Verb, non-3rd person singular present
- VBN – Verb, past participle
- VBG – Verb, gerund or present participle
- VBD – Verb, past tense

Implementation

```
% Python program

from nltk.tokenize import word_tokenize
from nltk import pos_tag

# Sample sentence
text = "POS tagging is essential for natural language
processing."

# Tokenize the sentence into words
words = word_tokenize(text)

# Perform part-of-speech tagging
pos_tags = pos_tag(words)

# Print the results
print("Original Text: \n", text)

print("\nPoS Tagging :")
for word, tag in pos_tags:
    print(f"{word} - {tag}")

% Output:

Original Text:
POS tagging is essential for natural language processing.
```

```
PoS Tagging :
POS - NNP
tagging - NN
is - VBZ
essential - JJ
for - IN
natural - JJ
language - NN
processing - NN
. - .
```

POS-tagging algorithms are categorized into two major types [2].

- Rule- based POS taggers
- Stochastic POS taggers

2.1.1 Rule-Based Tagging

Rule-based part-of-speech tagging [3] is assigning words their respective parts of speech using predefined linguistic rules. These rules are based on linguistic patterns, morphological features, and syntactic structures. It employs contextual information to provide tags to unfamiliar or confusing terms. Disambiguation is achieved by examining the grammatical characteristics of the word, its antecedent, its subsequent word, and additional factors. Like for example, rules like:

- If a word ends in "-ing," it is likely a verb.
- If a word starts with a capital letter and is not at the beginning of a sentence, it is likely a proper noun.
- Determiners (e.g., "the," "a") are often followed by nouns.

2.1.2 Stochastic POS Tagging

Stochastic POS taggers [3] use statistical models and machine learning techniques to learn the probability distribution of words given their POS tags. These models are trained on large annotated corpora, where words are labeled with their correct POS tags. Stochastic models can generalize well to handle unseen words or contexts. The most basic stochastic taggers resolve word ambiguity exclusively based on the likelihood of a word being associated with a specific tag. The tag most commonly associated with the word in the training set is applied to an ambiguous instance of that word.

Implementation

```
% Python program

import nltk
from nltk.tokenize import word_tokenize
from nltk.tag import RegexpTagger, hmm
from nltk.corpus import brown

# Sample sentence
sentence = "Part-of-speech tagging is essential for
natural language processing."

# Tokenize the sentence into words
words = word_tokenize(sentence)

# Rule-Based POS Tagger
def rule_based_tagger(words):
    rules = [
        (r'.*ed$', 'VBD'),   # Past tense verbs
        (r'.*ing$', 'VBG'),  # Gerunds
        (r'[A-Z][a-z]*$', 'NNP'),  # Proper nouns
        (r'.*$', 'NN')   # Default to noun
    ]

    regex_tagger = RegexpTagger(rules)
    return regex_tagger.tag(words)

# Stochastic (HMM) POS Tagger
def hmm_tagger(words):
    trainer = hmm.HiddenMarkovModelTrainer()
    model = trainer.train(nltk.corpus.brown.
tagged_sents())
    return model.tag(words)

# Apply both taggers
rule_based_tags = rule_based_tagger(words)
hmm_tags = hmm_tagger(words)

# Display the results
print("Rule-Based POS Tags:", rule_based_tags)
print("\nHMM POS Tags:", hmm_tags)
```

```
% Output:
```

```
Rule-Based POS Tags: [('Part-of-speech', 'NN'),
('tagging', 'VBG'), ('is', 'NN'), ('essential', 'NN'),
('for', 'NN'), ('natural', 'NN'), ('language', 'NN'),
('processing', 'VBG'), ('.', 'NN')]
```

```
HMM POS Tags: [('Part-of-speech', 'AT'), ('tagging',
'AT'), ('is', 'AT'), ('essential', 'AT'), ('for',
'AT'), ('natural', 'AT'), ('language', 'AT'),
('processing', 'AT'), ('.', 'AT')]
```

2.2 Stop Words

Stop words [4], are the words that are frequently occurring, are words that are programmed to ignore intentionally by search engine as they are perceived to make a minimal contribution to overall comprehension of a text.This exclusion occurs during indexing entries for searching and when presenting search results in response to a query.

Examples include "the," "is," "and," "in," etc.

```
% Python program
```

```
print(stopwords.words('english')[:20])
```

```
% Output:
```

```
['a', 'about', 'above', 'after', 'again', 'against',
'ain', 'all', 'am', 'an', 'and', 'any', 'are', 'aren',
"aren't", 'as', 'at', 'be', 'because', 'been']
```

2.3 Sequence Labeling

Sequence labeling is a natural language processing task that involves assigning a label or category to each element in a sequence of input data. In machine learning, this involves pattern recognition.

Two fundamental sequence labeling algorithms are the generative Hidden Markov Model (HMM) and the discriminative Conditional Random Field (CRF) [5]. For example, in part-of-speech tagging, each word in a sentence is designated with its appropriate part of speech.

2.3.1 Hidden Markov Model

The Hidden Markov Model (HMM) [6] is a statistical model widely used in various fields, including NLP. HMM is particularly effective for modeling sequences of observations and the underlying hidden states that generate them. A Markov chain is a mathematical model that illustrates a process in which the system shifts from one state to another. The transition posits that the likelihood of progressing to the subsequent state is exclusively contingent upon the present condition. The HMM is an extension of the Markov process utilized to represent phenomena in which the states are latent or hidden, yet nevertheless produce observable outputs. An HMM consists of two main components: a set of hidden states and a set of observable outputs or emissions.

- **Hidden States:** The system has a set of hidden states, and at any given time, it is in one of these states. However, the actual state is hidden or not directly observable.
- **Observable States (Emissions):** Each hidden state emits an observable output or observation.

In HMM taggers, probabilities are determined using maximum likelihood estimation applied to tag-labeled training corpora. The Viterbi algorithm is employed for decoding to identify the most probable tag sequence. HMM tagging is a generative approach.

Implementation

```
% Python program

import nltk
from nltk.tag import HiddenMarkovModelTrainer
from nltk.tokenize import word_tokenize

# Training data
training_data = [
    [('POS', 'NN'), ('tagging', 'VBG'), ('is', 'VBZ'),
('essential', 'JJ'),
     ('for', 'IN'), ('natural', 'JJ'), ('language',
'NN')],
    [('language', 'NN'), ('processing', 'VBG'), ('is',
'VBZ'), ('important', 'JJ')]
]
```

```
# Initialize and train the HMM model
trainer = HiddenMarkovModelTrainer()
hmm_model = trainer.train(training_data)

# Test sentence
test_sentence = "POS tagging is essential for natural
language processing."

# Tokenize the test sentence
tokens = word_tokenize(test_sentence)

# Perform POS tagging using the trained model
pos_tags = hmm_model.tag(tokens)

# Print the POS tags
print("Test Sentence:", test_sentence)
print("POS Tags:", pos_tags)

% Output:

Test Sentence: POS tagging is essential for natural
language processing.
POS Tags: [('POS', 'NN'), ('tagging', 'VBG'), ('is',
'VBZ'), ('essential', 'JJ'), ('for', 'IN'),
('natural', 'JJ'), ('language', 'NN'), ('processing',
'NN'), ('.', 'NN')]
```

2.3.2 *The Conditional Random Field*

Conditional Random Fields (CRFs) [7] are a type of probabilistic graphical model frequently applied in NLP and computer vision. They are a variation of Markov Random Fields (MRFs), a type of undirected graphical model.CRFs excel in structured prediction tasks, such as POS tagging in NLP, where the objective is to predict a structured output based on input features. Tasks like Named Entity Recognition (NER) and chunking also benefit from CRFs, especially when the output is a structured sequence. CRFs undergo training through maximum likelihood estimation, optimizing model parameters to maximize the probability of the correct output sequence given input features. Common optimization algorithms like gradient descent or L-BFGS are employed for solving this optimization problem. CRF tagging is a discriminative approach.

Implementation

```
% Python program

from sklearn_crfsuite import CRF
from sklearn_crfsuite.metrics import flat_accuracy_score
import nltk

# Sample Training Data (word, POS tag)
training_data = [
    [('POS', 'NN'), ('tagging', 'VBG'), ('is', 'VBZ'),
('essential', 'JJ'),
    ('for', 'IN'), ('natural', 'JJ'), ('language',
'NN')],
]

# Feature extraction function for CRF
def word2features(sent, i):
    word = sent[i][0]
    features = {
        'bias': 1.0,
        'word.lower()': word.lower(),
        'word[-3:]': word[-3:],
        'word.istitle()': word.istitle(),
        'word.isupper()': word.isupper(),
        'word.isdigit()': word.isdigit(),
    }

    if i > 0:  # Previous word
        prev_word = sent[i-1][0]
        features.update({
            '-1:word.lower()': prev_word.lower(),
            '-1:word.istitle()': prev_word.istitle(),
            '-1:word.isupper()': prev_word.isupper(),
        })
    else:
        features['BOS'] = True  # Beginning of sentence

    if i < len(sent)-1:  # Next word
        next_word = sent[i+1][0]
        features.update({
            '+1:word.lower()': next_word.lower(),
            '+1:word.istitle()': next_word.istitle(),
            '+1:word.isupper()': next_word.isupper(),
        })
```

```python
    else:
        features['EOS'] = True   # End of sentence

    return features

# Convert sentences into feature dictionaries
def sent2features(sent):
    return [word2features(sent, i) for i in
range(len(sent))]

# Convert sentences into labels (POS tags)
def sent2labels(sent):
    return [label for word, label in sent]

# Extract features and labels for training
X_train = [sent2features(sent) for sent in
training_data]
y_train = [sent2labels(sent) for sent in
training_data]

# Train the CRF model
crf_model = CRF(algorithm='lbfgs', max_iterations=100,
c1=0.1, c2=0.1)
crf_model.fit(X_train, y_train)

# Test sentence
test_sentence = "POS tagging is essential for natural
language processing."

# Tokenize the test sentence
tokens = nltk.word_tokenize(test_sentence)

# Extract features for the test sentence
test_features = [word2features([(token, '')], 0) for
token in tokens]

# Perform POS tagging using the trained CRF model
predicted_tags = crf_model.predict([test_features])[0]

# Print the results
print("Test Sentence:", test_sentence)
print("Predicted POS Tags:", predicted_tags)
for token, tag in zip(tokens, predicted_tags):
    print(f"{token}: {tag}")
```

```
% Output:

Test Sentence: POS tagging is essential for natural
language processing.
Predicted POS Tags: ['NN' 'VBG' 'VBZ' 'JJ' 'IN' 'JJ'
'NN' 'VBG' 'NN']
POS: NN
tagging: VBG
is: VBZ
essential: JJ
for: IN
natural: JJ
language: NN
processing: VBG
.: NN
```

2.4 Context-Free Grammar (CFG)

Grammar is defined as the rules for forming well-structured sentences [8]. Grammar is crucial in outlining the syntactic structure of well-formed programs, serving as a set of rules. In simpler terms, Grammar denotes syntactical rules that are used for conversation in natural languages.

Grammar G can be written as a 4-tuple (N, T, S, P) where,

- N or V = set of non-terminal symbols, or variables.
- T or Σ = set of terminal symbols.
- P = Production rules for Terminals as well as Non-terminals.
- S = Start symbol where S ∈ N

Set of Non-terminals (V): Syntactic variables representing sets of strings, defining the language generated by the grammar.

Set of Terminals (Σ): Also called tokens, these are basic symbols used to form strings.

Set of Productions (P): Specifies how non-terminals and terminals can be combined, each production having non-terminals, an arrow, and a sequence of terminals.

Start Symbol (S): The production starts with the designated start symbol "S," typically a non-terminal symbol. Non-terminals are always assigned as start symbols.

A context-free grammar, abbreviated as CFG, is a notation utilized for defining languages and serves as a superset of regular grammar.

$G \rightarrow (V \cup T)^*$, where G ∈ V

2.5 Parsing

Parsing is a fundamental step in NLP where a sentence is analyzed to identify its grammatical structure and its underlying structure to extract meaning from it. This process involves breaking down the sentence into smaller components to extract meaning, enabling machines to understand human language by examining the syntax and underlying structure of the text. It is analyzing the input sentence by breaking it down into its grammatical constituents, identifying the parts of speech, and syntactic relations. Parsing with CFGs includes the assignment of appropriate parse trees to input strings, where the tree encompasses all and only the constituents of the input, culminating in a root labeled "S."

2.5.1 Types of Parsing

- **Top – Down Parsing**: Starts from the root (sentence) and recursively breaks it down into smaller constituents.
- **Bottom-up Parsing**: Starts with words (tokens) and builds up to the full sentence structure.

```
% Python program

import nltk

# Define the grammar
grammar = nltk.CFG.fromstring("""
S -> NP VP
NP -> Det N | 'John'
VP -> V NP
Det -> 'the'
N -> 'man'
V -> 'saw'
""")

# Sentence to parse
sentence = ['the', 'man', 'saw', 'John']

top_down_parser = nltk.RecursiveDescentParser(grammar)

# Parse the sentence and print the tree
for tree in top_down_parser.parse(sentence):
    tree.pretty_print()
```

```
                    S
          _____|_____
         NP                   VP
        __|__                __|__
      Det    N             V      NP
       |     |             |      |
      the   man           saw    John
```

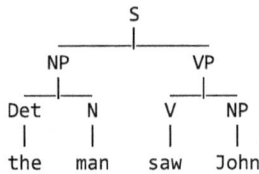

% Python program

```python
import nltk

# Define the grammar
grammar = nltk.CFG.fromstring("""
S -> NP VP
NP -> Det N | 'John'
VP -> V NP
Det -> 'the'
N -> 'man'
V -> 'saw'
""")

# Sentence to parse
sentence = ['the', 'man', 'saw', 'John']

bottom_down_parser  = nltk.ChartParser(grammar)

# Parse the sentence and print the tree
for tree in bottom_down_parser.parse(sentence):
    tree.pretty_print()
```

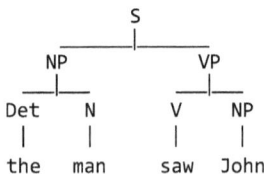

% Output:

```
                    S
          _____|_____
         NP                   VP
        __|__                __|__
      Det    N             V      NP
       |     |             |      |
      the   man           saw    John
```

2.5.2 Earley Parsing

Earley parsing [9] is an efficient top-down parsing algorithm that miti-
gates some inefficiencies inherent in a merely naïve search employing the

same top-down approach. Intermediate answers are generated singularly and recorded in a chart (dynamic programming). The issue of left recursion is addressed by the analysis of the input. Earley is not selective regarding the sort of grammar it accommodates; it accepts any context-free grammar and populates a table in a single pass over the input.

Steps:

1. Predict sub-structure (based on grammar)
2. Scan partial solutions for a match
3. Complete a sub-structure (i.e., build constituents)

Implementation

```
% Python program

import nltk

# Define the grammar
grammar = nltk.CFG.fromstring("""
S -> NP VP
NP -> Det N | 'John'
VP -> V NP
Det -> 'the'
N -> 'man'
V -> 'saw'
""")

# Sentence to parse
sentence = ['the', 'man', 'saw', 'John']

# Create the EarleyChartParser
earley_parser  = nltk.EarleyChartParser(grammar)

# Parse the sentence and print the tree
for tree in earley_parser.parse(sentence):
    tree.pretty_print()

% Output:
```

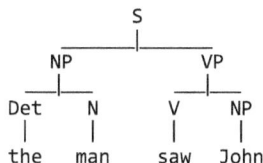

```
              S
      _____|_____
     NP               VP
    __|__            __|__
   Det   N          V    NP
    |    |          |    |
   the  man        saw  John
```

2.5.3 Cocke-Kasami-Younger Parsing

The Cocke-Kasami-Younger (CKY) algorithm [10], the most widely used dynamic-programming-based approach to parsing, is implemented using a matrix to keep track of partial results. The algorithm operates on the assumption that the solution to issue [i, j] can be derived from the solutions to the subproblems [i, k] and [k, j]. The method necessitates that the Grammar G be in Chomsky Normal Form (CNF). CNF consists of rules that contain either two non-terminals or a terminal on the right-hand side. Efficient parsing algorithm utilizing tabulation of substring parses to eliminate redundant computations.

Steps of the CYK algorithm:

1. Initialization of the parse table.
2. Filling in the parse table by considering all possible productions.
3. Determining whether the start symbol is derivable from the input string.

Implementation

```
% Python program

import nltk

# Define the grammar
grammar = nltk.CFG.fromstring("""
S -> NP VP
NP -> Det N | 'John'
VP -> V NP
Det -> 'the'
N -> 'man'
V -> 'saw'
""")

# Sentence to parse
sentence = ['the', 'man', 'saw', 'John']

cky_parser = nltk.ChartParser(grammar)

# Parse the sentence and print the tree
for tree in cky_parser.parse(sentence):
    tree.pretty_print()
```

% Output:

```
                    S
            _____|_____
          NP              VP
         _|_             _|_
       Det   N          V    NP
        |    |          |    |
       the  man        saw  John
```

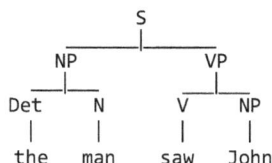

2.6 Probabilistic Context-Free Grammar

Probabilistic Context-Free Grammars (PCFGs) are an extension of traditional CFGs used in the field of computational linguistics and natural language processing. While CFGs describe the syntactic structure of languages through a set of production rules, PCFGs enhance this framework by associating probabilities with each production rule.

Mathematically, the PCFG can be represented as:
$G = (N, \Sigma, P, S, R)$ where

- N is the set of non-terminals.
- is the set of terminals.
- P is the set of production rules.
- S is the start symbol.
- R is the set of probabilities associated with each production rule.

A production rule with probability can be represented as
$A \rightarrow \beta[p]$, indicating that the rule $A \rightarrow \beta$ has a probability p
Steps:

1. Define a set of context-free grammar production rules.
2. Assign probabilities to each production rule based on observed frequencies in a training corpus.
3. Ensure that the probabilities associated with all rules for a given non-terminal sum to 1.
4. Use the PCFG for parsing by selecting production rules probabilistically during derivation.
5. Apply the Viterbi algorithm to find the most likely parse tree, considering rule probabilities.

6. Generate multiple possible parse trees for a sentence, each with an associated probability.
7. Train the PCFG by estimating probabilities from annotated linguistic data

Implementation

```
% Python program

import nltk

# Define a Probabilistic Context-Free Grammar (PCFG)
pcfg_grammar = nltk.PCFG.fromstring("""
S -> NP VP [1.0]
NP -> Det N [0.7] | 'John' [0.3]
VP -> V NP [0.9] | VP PP [0.1]
Det -> 'the' [0.8] | 'a' [0.2]
N -> 'man' [0.5] | 'dog' [0.5]
V -> 'chased' [0.6] | 'saw' [0.4]
PP -> P NP [1.0]
P -> 'in' [0.4] | 'on' [0.6]
""")

# Input sentence
sentence = ['the', 'man', 'saw', 'John']

# Create a PCFG parser
pcfg_parser = nltk.ViterbiParser(pcfg_grammar)

# Parse the sentence
for tree in pcfg_parser.parse(sentence):

    tree.pretty_print()


% Output:
```

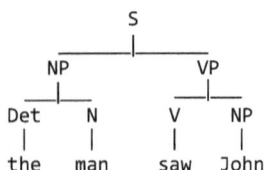

2.7 Term Frequency and Inverse Document Frequency

The Term Frequency and Inverse Document Frequency (TF-IDF) [11,12] representation takes into account the importance of each word in a document. It is a numerical statistic that reflects the importance of a word in a document relative to its occurrence across a collection of documents.

Term Frequency (TF): TF quantifies the occurrence of a term within a document.

$$\mathrm{TF}(t,d) = \frac{\text{Total number of terms in document d}}{\text{Number of times term t appears in document d}}$$

Inverse Document Frequency (IDF): IDF quantifies the infrequency of a phrase within a corpus of documents.

$$\mathrm{IDF}(t,D) = \log\left(\frac{\text{Number of documents containing term } t+1}{\text{Total number of documents in the collection } |D|}\right)$$

TF-IDF Calculation: TF-IDF is derived by multiplying the TF and IDF values for a certain term within a document.

$$\mathrm{TF} - \mathrm{IDF}(t,d,D) = \mathrm{TF}(t,d) \times \mathrm{IDF}(t,D)$$

where

- D is the collection of documents,
- d is the document,
- t is the term (word).

Steps:

1. Data Pre-processing
2. Calculating Term Frequency
3. Calculating Inverse Document Frequency
4. Calculating Product of Term Frequency & Inverse Document Frequency

Implementation

```
% Python program

import pandas as pd
```

```python
from sklearn.feature_extraction.text import
TfidfVectorizer

# Assign documents
d0 = 'Natural language processing is fascinating'
d1 = 'Processing language using Python'
d2 = 'Python is widely used in data science'

# Merge documents into a single corpus
documents = [d0, d1, d2]

# Create TfidfVectorizer object
tfidf = TfidfVectorizer()

# Get TF-IDF values
result = tfidf.fit_transform(documents)

# Create DataFrame for IDF values
idf_df = pd.DataFrame(data={"Term": tfidf.get_feature_
names_out(), 'IDF': tfidf.idf_})

# Create DataFrame for TF-IDF matrix
tfidf_matrix_df = pd.DataFrame(data=result.toarray(),
columns=tfidf.get_feature_names_out())

# Display IDF values
print("\nIDF Values:")
print(idf_df)

# Display TF-IDF Matrix
print("\nTF-IDF Matrix:")
print(tfidf_matrix_df)
```

```
% Output:


IDF Values:
          Term      IDF
0         data 1.693147
1   fascinating 1.693147
2           in 1.693147
3           is 1.287682
4     language 1.287682
5      natural 1.693147
6   processing 1.287682
7       python 1.287682
8      science 1.693147
9         used 1.693147
10       using 1.693147
11      widely 1.693147


TF-IDF Matrix:
        data  fascinating       in       is  language  natural  processing  \
0   0.000000      0.51742 0.000000 0.393511  0.393511  0.51742    0.393511
1   0.000000      0.00000 0.000000 0.000000  0.459854  0.00000    0.459854
2   0.403016      0.00000 0.403016 0.306504  0.000000  0.00000    0.000000


      python   science     used    using    widely
0   0.000000  0.000000 0.000000 0.000000  0.000000
1   0.459854  0.000000 0.000000 0.604652  0.000000
2   0.306504  0.403016 0.403016 0.000000  0.403016
```

2.8 Information Extraction

Information extraction involves converting unstructured data into organized, editable formats. This can include identifying entities (such as names, locations, organizations) and extracting relevant information associated with those entities.

Information Extraction Systems find applications in diverse areas, from summarizing vast text collections to powering conversational AI systems and virtual assistants like Apple's Siri, Amazon's Alexa, and, Google Assistant highlighting their reliance on sophisticated IE systems.

Implementation –

```
% Python program

import spacy

# Load the English NLP model from spaCy
nlp = spacy.load("en_core_web_sm")
```

```
# Sample text for information extraction
text = "Apple Inc. is a technology company based in
Cupertino, California. It was founded by Steve Jobs,
Steve Wozniak, and Ronald Wayne in 1976."

# Process the text using spaCy
doc = nlp(text)

# Extract entities (names, locations, organizations)
entities = [(ent.text, ent.label_) for ent in doc.ents]

# Print extracted entities
print("Entities:", entities)

% Output:

Entities: [('Apple Inc.', 'ORG'), ('Cupertino', 'GPE'),
('California', 'GPE'), ('Steve Jobs', 'PERSON'),
('Steve Wozniak', 'PERSON'), ('Ronald Wayne',
'PERSON'), ('1976', 'DATE')]
```

2.9 Relation Extraction

Relation extraction [13] involves predicting attributes and relationships for entities within a sentence. Relation extraction involves identifying and classifying relationships between entities mentioned in the text.

This task is fundamental for constructing relation knowledge graphs and holds immense importance in NLP applications, including summarization, question answering, sentiment analysis, and structured search.

Implementation

```
% Python program

import spacy

# Load the spacy English NLP model
nlp = spacy.load("en_core_web_sm")

# Sample text for relation extraction
text = "Steve Jobs was one of the founders of Apple Inc."
```

```
# Process the text using spacy
doc = nlp(text)

# Extract relations (founder relationship)
relations = []

for ent in doc.ents:
    if ent.label_ == 'ORG':  # Checking for
organization entity
        org_name = ent.text
    elif ent.label_ == 'PERSON' : # Checking for
person entity and making sure org_name is set
        founder_name = ent.text

relations.append((founder_name, 'founder', org_name))

# Print the relations
print("Relations:", relations)

# Print each entity found in the text
print("\nEntities found:")
for ent in doc.ents:
    print((ent.text, ent.label_))

% Output:

Relations: [('Steve Jobs', 'founder', 'Apple Inc.')]
Entities found:
('Steve Jobs', 'PERSON')
('Apple Inc.', 'ORG')
```

2.10 Summary

This chapter explores syntactic analysis, a crucial aspect of NLP that focuses on understanding the grammatical structure of sentences. It begins with POS tagging, covering both rule-based and stochastic approaches. The role of stop words in text processing is discussed, along with sequence labeling techniques such as HMM and CRF, which help in tagging words based on contextual dependencies. This chapter then introduces CFG and parsing techniques, including top-down parsing, Earley parsing, and CKY parsing, which are essential for syntactic structure identification.

The concept of PCFG is explored to incorporate statistical probabilities into grammatical structures. Additionally, this chapter covers TF-IDF, a widely used technique in information retrieval and text mining. Finally, it discusses information extraction and relation extraction, which are critical for identifying structured data from unstructured text. Overall, this chapter provides a comprehensive understanding of how NLP models analyze and interpret sentence structures to extract meaningful information.

References

1. A. R. Martinez, "Part-of-speech tagging," *Wiley Interdiscip. Rev. Comput. Stat.*, vol. 4, no. 1, pp. 107–113, 2012.
2. J. Kupiec, "Robust part-of-speech tagging using a hidden Markov model," *Comput. Speech Lang.*, vol. 6, no. 3, pp. 225–242, 1992.
3. D. Kumawat and V. Jain, "POS tagging approaches: A comparison," *Int. J. Comput. Appl.*, vol. 118, no. 6, 2015.
4. W. J. Wilbur and K. Sirotkin, "The automatic identification of stop words," *J. Inf. Sci.*, vol. 18, no. 1, pp. 45–55, 1992.
5. J. Lafferty, A. McCallum, F. Pereira, and others, "Conditional random fields: Probabilistic models for segmenting and labeling sequence data," in International Conference on Machine Learning (ICML 2001), University of New South Wales, USA, 2001, p. 3.
6. V. P. Chandrika, R. Verma, N. Charan, S. Ditheswar, S. Hansika, and R. Ishwariya, "POS Tagging Using Hidden Markov Models in Natural Language Processing," in *2024 International Conference on Signal Processing, Computation, Electronics, Power and Telecommunication (IConSCEPT)*, Puducherry Karaikal, India, 2024, pp. 1–6.
7. C. Sutton, A. McCallum, and others, "An introduction to conditional random fields," *Found. Trends® in Mach. Learn.*, vol. 4, no. 4, pp. 267–373, 2012.
8. A. McCallum, "Context Free Grammars," 2004.
9. J. Aycock and R. N. Horspool, "Practical earley parsing," *Comput. J.*, vol. 45, no. 6, pp. 620–630, 2002.
10. M. Nederhof and G. Satta, "Theory of parsing," *Handb. Comput. Linguist. Nat. Lang. Process.*, vol. 82, pp. 105–130, 2010.
11. G. Salton, *Modern Information Retrieval*, McGraw-Hill, 1983.
12. W. I. D. Mining, "Data mining: Concepts and techniques," *Morgan Kaufinann*, vol. 10, no. 559–569, p. 4, 2006.
13. N. Konstantinova, "Review of relation extraction methods: What is new out there?," in *Analysis of Images, Social Networks and Texts: Third International Conference, AIST 2014*, Yekaterinburg, Russia, April 10–12, 2014, Revised Selected Papers 3, 2014, pp. 15–28.

3

SEMANTIC ANALYSIS

In order to assist machines understand meaning, semantic analysis extracts meaning from language and establishes the framework for a semantic system. Lexical, grammatical, and syntactic analysis are all used in semantic analysis to decipher sentence structure and determine word meanings so that machines can comprehend language just as well as people. Lexical semantics, which examines the meanings of individual words (i.e., dictionary definitions), is the first step in semantic analysis. The meaning of words that combine to make a phrase is then examined using semantic analysis, which also looks at the links between individual words. Words in context are clearly understood thanks to this analysis. For text analysis to be very accurate, semantic analysis is essential.

Example – "He used the bat to hit a home run."

In this context, "bat" refers to a piece of sports equipment used in games like baseball or cricket.

-"I saw a bat flying in the night sky."

Here, "bat" refers to the mammal capable of flight.

3.1 Semantic Grammar

The collection of guidelines and precepts that control the meaning of linguistic constructions is known as semantic grammar. The contribution of words and phrases to the overall meaning of a sentence or speech is the focus of these grammars. Semantic grammars explore the meaning of language elements, as opposed to syntactic grammars, which concentrate on the form of sentences. In computational linguistics and NLP, semantic grammars are essential. They enable programs like sentiment analysis, language translation, and question-answering systems by assisting computers in comprehending the intended meaning of utterances.

DOI: 10.1201/9781003425328-3   **49**

Implementation

```
% Python program

import nltk

# Define a simple semantic grammar
semantic_grammar = nltk.CFG.fromstring("""
S -> NP VP
NP -> Det N
VP -> V NP
Det -> 'the'
N -> 'cat' | 'dog'
V -> 'chased' | 'caught'
""")

# Parse a sentence using the defined grammar
sentence = "the cat chased the dog".split()

# Create a parser based on the semantic grammar
parser = nltk.ChartParser(semantic_grammar)

# Parse the sentence and print the tree
for tree in parser.parse(sentence):
    tree.pretty_print()

% Output:
```

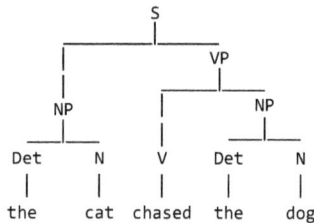

```
                       S
        _____|___
       |                   VP
       |                ___|_____
       NP              |            NP
    ___|___            |         ___|___
   Det      N          V        Det      N
    |       |          |         |       |
   the     cat      chased      the     dog
```

3.2 Lexical Semantics

It constitutes the initial phase of semantic analysis, where the focus is on comprehending the meanings of individual words [1]. This process encompasses the examination of words, sub-words, affixes (sub-units), compound words, and phrases. Collectively, these elements are referred to as lexical items. In simpler terms, lexical semantics explores the connections among lexical items, the meaning conveyed by sentences, and the syntax employed in constructing those sentences.

Implementation

```
% Python program

from nltk.corpus import wordnet

synonyms = []
antonyms = []

for syn in wordnet.synsets("happy"):
    for lemma in syn.lemmas():
        synonyms.append(lemma.name())
        if lemma.antonyms():
            antonyms.append(lemma.antonyms()[0].name())

print("Synonyms:", set(synonyms))
print("Antonyms:", set(antonyms))

% Output:

Synonyms: {'well-chosen', 'glad', 'happy', 'felicitous'}
Antonyms: {'unhappy'}
```

3.3 Lexemes

All variations of a single word are represented by a lemma, a fundamental word form. With a part-of-speech and a collection of associated word senses, a lexeme [2] is an abstract representation of a word (as well as all of its forms). (Usually simply written (or called) the lemma, maybe using a different font). Lexemes are meaning units that frequently match words or stems. The study of lexical semantics looks at the connections between words and their different meanings.

3.4 Word Senses

A sense refers to a specific meaning or interpretation of a word in a given context [3]. Understanding these relationships is essential for capturing the richness of meaning in natural language and developing accurate semantic models for computational applications.

3.4.1 Hyponymy

It illustrates how a generic term and its examples relate to one another. In this context, a hypernym is a generic phrase, and hyponyms are examples of it [4].

Example- "Rose" is a hyponym of the hypernym "Flower" since a rose is a specific type of flower.

- The word color is hypernym, and the colors blue, yellow, green, etc. are hyponyms.

```
% Python program

from nltk.corpus import wordnet

hyponyms_color= wordnet.synset('color.n.01').
hyponyms()

print("Hyponyms of 'color':", hyponyms_color)

% Output:

Hyponyms of 'color': [Synset('coloration.n.02'),
Synset('chromatic_color.n.01'),
Synset('complexion.n.01'), Synset('mottle.n.01'),
Synset('achromatic_color.n.01'), Synset('nonsolid_
color.n.01'), Synset('primary_color.n.01'),
Synset('shade.n.02'), Synset('heather_mixture.n.01')]
```

3.4.2 Homonymy

More than one lexical terms that have distinct meanings yet the same spelling. It can be characterized as a group of words with similar spellings or forms but distinct and unconnected meanings. Example- "Bat" can refer to a sports equipment used in baseball or a flying mammal.

```
% Python program

from nltk.corpus import wordnet

homonym1 = wordnet.synsets('bat') # Flying mammal vs.
sports equipment
```

```
homonym2 = wordnet.synsets('bank') # Financial
institution vs. river bank

print("Homonym 1:", homonym1)
print("\nHomonym 2:", homonym2)

% Output:

Homonym 1: [Synset('bat.n.01'), Synset('bat.n.02'),
Synset('squash_racket.n.01'), Synset('cricket_
bat.n.01'), Synset('bat.n.05'), Synset('bat.v.01'),
Synset('bat.v.02'), Synset('bat.v.03'),
Synset('bat.v.04'), Synset('cream.v.02')]

Homonym 2: [Synset('bank.n.01'), Synset('depository_
financial_institution.n.01'), Synset('bank.n.03'),
Synset('bank.n.04'), Synset('bank.n.05'),
Synset('bank.n.06'), Synset('bank.n.07'),
Synset('savings_bank.n.02'), Synset('bank.n.09'),
Synset('bank.n.10'), Synset('bank.v.01'),
Synset('bank.v.02'), Synset('bank.v.03'),
Synset('bank.v.04'), Synset('bank.v.05'),
Synset('deposit.v.02'), Synset('bank.v.07'),
Synset('trust.v.01')]
```

3.4.3 Polysemy

It is a term or expression with a distinct but connected meaning. To put it another way, polysemy [5] is the use of the same spelling with related but distinct meanings. Example- "Run" can mean to move swiftly (e.g., "She runs every morning") or to operate something (e.g., "to run a machine").

```
% Python program

from nltk.corpus import wordnet

polysemy = wordnet.synsets('bank') # Different senses
related to finance

print("Polysemy:", polysemy)
```

```
% Output:

Polysemy: [Synset('bank.n.01'), Synset('depository_
financial_institution.n.01'), Synset('bank.n.03'),
Synset('bank.n.04'), Synset('bank.n.05'),
Synset('bank.n.06'), Synset('bank.n.07'),
Synset('savings_bank.n.02'), Synset('bank.n.09'),
Synset('bank.n.10'), Synset('bank.v.01'),
Synset('bank.v.02'), Synset('bank.v.03'),
Synset('bank.v.04'), Synset('bank.v.05'),
Synset('deposit.v.02'), Synset('bank.v.07'),
Synset('trust.v.01')]
```

3.4.4 Synonymy

A minimum of two lexical terms that have similar meanings but distinct spellings. Synonymy [6] refers to the relationship between two or more words that have different spellings but share similar meanings. These words are interchangeable in certain contexts. Example – Big/Large, Begin/Commence, author/writer, fate/destiny

```
% Python program

from nltk.corpus import wordnet

synonyms = wordnet.synsets('happy') # Synonyms of
'happy'

print("Synonyms:", synonyms)

% Output:

Synonyms: [Synset('happy.a.01'),
Synset('felicitous.s.02'), Synset('glad.s.02'),
Synset('happy.s.04')]
```

3.4.5 Antonymy

A pair of terms that have different meanings. A linguistic link between two words with opposing meanings is known as an antonym. The words in an antonymous pair are opposites. Example – Hot/Cold, Happy/Sad, life/death, moon/sun

```
% Python program

from nltk.corpus import wordnet

antonyms = []

for syn in wordnet.synsets("happy"):
  for lemma in syn.lemmas():
    if lemma.antonyms():
      antonyms.append(lemma.antonyms()[0].name())
print("Antonym", set(antonyms))

% Output:

Antonym {'unhappy'}
```

3.5 Wordnet

WordNet [7] organizes words into synsets, which are collections of synonyms that each stand for a different idea. It facilitates comprehension of word significance and relationships by capturing different semantic interactions between words. According to Fellbaum (1998), the WordNet lexical database is the main source for English sense relations. With the exception of closed-class terms, WordNet consists

WordNet Search - 3.0 - WordNet home page - Glossary - Help

Word to search for: bass   (Search WordNet)

Display Options: (Select option to change) [:] (Change)

Key: "S:" = Show Synset (semantic) relations, "W:" = Show Word (lexical) relations

Noun

- S: (n) **bass** (the lowest part of the musical range)
- S: (n) **bass**, bass part (the lowest part in polyphonic music)
- S: (n) **bass**, basso (an adult male singer with the lowest voice)
- S: (n) sea bass, **bass** (the lean flesh of a saltwater fish of the family Serranidae)
- S: (n) freshwater bass, **bass** (any of various North American freshwater fish with lean flesh (especially of the genus Micropterus))
- S: (n) **bass**, bass voice, basso (the lowest adult male singing voice)
- S: (n) **bass** (the member with the lowest range of a family of musical instruments)
- S: (n) **bass** (nontechnical name for any of numerous edible marine and freshwater spiny-finned fishes)

Adjective

- S: (adj) **bass**, deep (having or denoting a low vocal or instrumental range) *"a deep voice"; "a bass voice is lower than a baritone voice"; "a bass clarinet"*

WordNet home page

of three separate databases for nouns, verbs, and adjectives/adverbs. Every database has lemmas with sense annotations. WordNet 3.0 has 4,481 adverbs, 22,479 adjectives, 11,529 verbs, and 117,798 nouns. Verbs typically have 2.16 senses, while nouns typically have 1.23 senses. WordNet is accessible online or can be downloaded for local access.

3.6 Word Similarity

Word similarity [8] measures the degree of closeness or relatedness between two words in terms of meaning. It's crucial for tasks like information retrieval, machine translation, and natural language processing.

The WordNet network's word distances allow us to calculate the similarity between two words. The terms are more similar the closer they are to one another. In this manner, it is possible to determine statistically that a phone and a computer are comparable, a cat and a dog are identical, yet a cat and a phone are not!

```
% Python program

from nltk.corpus import wordnet

# Get the synsets for 'dog' and 'cat'
word1 = wordnet.synset('dog.n.01')
word2 = wordnet.synset('cat.n.01')

# Calculate the path similarity
similarity_score = word1.path_similarity(word2)

# Print the similarity score
print("Similarity between 'dog' and 'cat':",
similarity_score)

% Output:

Similarity between 'dog' and 'cat': 0.2
```

Similarly, it is possible to quantitatively figure out that a sun and moon are similar, a car and a bicycle are similar, but sun and a car are not similar!

```
% Python program

from nltk.corpus import wordnet

def word_similarity(word1, word2):
    # Get the synsets for the words
    synset1 = wordnet.synsets(word1)
    synset2 = wordnet.synsets(word2)

    if synset1 and synset2:
        # Consider the first synset of each word
        synset1 = synset1[0]
        synset2 = synset2[0]

        # Calculate path similarity
        similarity = synset1.path_similarity(synset2)

        # Return similarity score, or 0.0 if no
similarity found
        return similarity if similarity is not None
else 0.0
    else:
        # One or both words are not found in WordNet
        return 0.0

# Example Usage:
similarity_sun_moon = word_similarity("sun", "moon")
similarity_car_bicycle = word_similarity("car",
"bicycle")
similarity_sun_car = word_similarity("sun", "car")

# Print the similarity results
print(f"Similarity between 'sun' and 'moon':
{similarity_sun_moon}")
print(f"Similarity between 'car' and 'bicycle':
{similarity_car_bicycle}")
print(f"Similarity between 'sun' and 'car':
{similarity_sun_car}")

% Output:

Similarity between 'sun' and 'moon': 0.2
Similarity between 'car' and 'bicycle': 0.2
Similarity between 'sun' and 'car':
0.08333333333333333
```

3.7 Word Sense Disambiguation

Word Sense Disambiguation (WSD) [9] plays a crucial role in NLP by helping to figure out the meaning of a word within a specific context. NLP systems frequently encounter the challenge of accurately recognizing words, and deciphering the precise meaning of a word in a given sentence is applicable in various scenarios. WSD is like a detective for words in the NLP. A word has different meanings in different situations. For example, the word "bank" could mean a place where you keep money or the side of a river. When computers read a sentence, they might get confused about which meaning of a word is being used. WSD helps the computer figure out the correct meaning by looking at the context or the other words in the sentence.

There are four main ways to implement WSD.

1. Dictionary- and knowledge-based methods
2. Supervised methods
3. Semi-supervised methods
4. Unsupervised methods

Implementation

```
% Python program

from nltk.wsd import lesk
from nltk.tokenize import word_tokenize

# Sample sentence with an ambiguous word
sentence = "I went to the bank to deposit money."

# Ambiguous word
ambiguous_word = 'bank'

# Tokenize the sentence
tokenized_sentence = word_tokenize(sentence)

# Perform WSD using the Lesk algorithm
sense = lesk(tokenized_sentence, ambiguous_word)
```

```
# Print the disambiguated sense
print(f"Ambiguous Word: {ambiguous_word}")
print(f"Disambiguated Sense: {sense}")
% Output:

Ambiguous Word: bank
Disambiguated Sense: Synset('savings_bank.n.02')
```

3.7.1 Dictionary Based Approach of WSD

A Dictionary-Based Approach of WSD [10] involves using dictionaries or lexical resources to determine the correct meaning or sense of a word in a given context. This method relies on the definitions and senses provided in a dictionary to disambiguate words.

Steps:

1. **Context Extraction:** Identify the context words surrounding the target word.
2. **Sense Inventory:** Use a sense inventory from a dictionary or lexical resource.
3. **Matching:** Compare the context words with the definitions or senses to select the most appropriate sense.

3.7.1.1 Lesk Algorithm The popularly used Lesk method is a seminal dictionary-based method. The Lesk algorithm [11] is a method used for WSD, which aims to determine the correct meaning of a word in a given context. It was introduced by Michael Lesk in 1986 and is based on the idea of using the context of the surrounding words to disambiguate the sense of a target word.

Steps

1. **Tokenization:** The method of dividing a text into discrete words or tokens.
2. **Selection of Ambiguous Word:** Choosing the word for which the meaning needs to be disambiguated.
3. **Gather Context Words:** Collecting the set of words appearing in the local context of the ambiguous word.
4. **Retrieve Synsets:** Obtaining the possible meanings or senses (synsets) of the ambiguous word from a lexical resource.

5. **Compute Overlap:** Calculating the similarity between the words in the context and the words in the definitions of each synset.
6. **Select the Best Sense:** Choosing the sense with the highest overlap as the most likely meaning for the ambiguous word

Implementation

```
% Python program

from nltk.wsd import lesk
from nltk.tokenize import word_tokenize

# Example 1
context1 = "I need to deposit my money in the bank."
ambiguous_word1 = "bank"

result1 = lesk(word_tokenize(context1), ambiguous_word1)
print(result1, result1.definition())

# Example 2
context2 = "The boat is floating down the river bank."
ambiguous_word2 = "bank"

result2 = lesk(word_tokenize(context2), ambiguous_word2)
print(result2, result2.definition())

# Example 3
context3 = "I sat on the bank of the lake, enjoying
the sunset."
ambiguous_word3 = "bank"

result3 = lesk(word_tokenize(context3), ambiguous_word3)
print(result3, result3.definition())

% Output:

Synset('savings_bank.n.02') a container (usually with
a slot in the top) for keeping money at home
Synset('bank.n.07') a slope in the turn of a road or
track; the outside is higher than the inside in order
to reduce the effects of centrifugal force
Synset('bank.v.07') cover with ashes so to control the
rate of burning
```

3.8 Information Retrieval

Information Retrieval [12–14] in semantic analysis of NLP involves leveraging semantic understanding to enhance the accuracy and relevance of retrieving information from large textual datasets. It includes processing user queries, indexing documents with semantic information, matching queries with documents based on meaning, and ranking the results to present the most relevant information to the user.

It is efficient retrieval of relevant information from a large collection of documents or textual data based on user queries or information needs. It plays a crucial role in tasks like search engines, document retrieval, and information extraction.

Information Retrieval (IR) has a few components as the process aims to efficiently retrieve relevant information by incorporating semantic understanding into each stage of information retrieval.

1. **Query Processing:** Analyze user queries using semantic analysis techniques to understand the meaning and intent behind the input.
2. **Indexing:** Create an index of the textual data, mapping terms to their locations, and incorporate semantic analysis to capture the meaning of terms.
3. **Semantic Matching:** Match user queries with indexed documents, employing semantic analysis to understand semantic similarity, including synonyms and related concepts.
4. **Ranking:** Rank retrieved documents based on their relevance to the user query, utilizing semantic analysis to assign weights to terms and concepts for accurate ranking.

Example

```
% Python program

from sklearn.feature_extraction.text import
TfidfVectorizer
from sklearn.metrics.pairwise import cosine_similarity

# Sample documents in a dataset
documents = [
    "Semantic analysis involves understanding the
meaning of text.",
```

```
    "Information retrieval is the process of obtaining
information from a large dataset.",
    "The TfidfVectorizer is a tool for transforming
text data into a numerical format for analysis."
]

# User query
query = "What is semantic analysis in NLP?"

# Apply TF-IDF vectorization
vectorizer = TfidfVectorizer()
tfidf_matrix = vectorizer.fit_transform(documents)

# Transform the user query into a vector
query_vector = vectorizer.transform([query])

# Calculate cosine similarity between the query and
each document
cosine_similarities = cosine_similarity(query_vector,
tfidf_matrix)

# Find the most relevant document
most_relevant_document_index = cosine_similarities.
argmax()

# Print the most relevant document
print("User Query:", query)
print("Most Relevant Document:")
print(documents[most_relevant_document_index])

% Output:

User Query: What is semantic analysis in NLP?
Most Relevant Document:
Semantic analysis involves understanding the meaning
of text.
```

Information retrieval (IR) is a software program managing the organization, storage, and retrieval of textual information. It helps users locate information by indicating the existence and location of potentially relevant documents in repositories. The goal is to retrieve documents that meet user requirements, aiming for a system that exclusively returns relevant information.

3.9 Summary

This chapter delves into semantic analysis, a key aspect of NLP that focuses on understanding the meaning of words and their relationships within a language. It introduces semantic grammars and lexical semantics, which help in defining the structure and meaning of words in different contexts. Various linguistic phenomena, such as homonymy (words with multiple meanings), polysemy (words with related meanings), synonymy (words with similar meanings), and hyponymy (hierarchical word relationships), are explored to illustrate the complexity of word meaning. This chapter also discusses WordNet, a large lexical database that organizes words based on their semantic relationships, and word similarity measures, which are crucial for applications like information retrieval and text classification. A key focus is WSD, which helps in determining the correct meaning of a word in a given context. Both dictionary-based approaches and computational techniques for WSD are examined. Lastly, this chapter explores information retrieval, demonstrating how semantic analysis enhances search engines, document classification, and text mining by improving the accuracy of extracting relevant information from large text datasets.

References

1. C. Paradis, "Lexical semantics," in C.A. Chapelle (ed.) *The Encyclopedia of Applied Linguistics*, Wiley-Blackwell, Hoboken, NJ, 2012.
2. F. Nemo, "Morphemes and lexemes versus 'Morphemes or Lexemes?,'" in *Proceeding of the Mediterranean Morphology Meetings*, Catania, 2003, pp. 195–208.
3. D. McCarthy, "Word sense disambiguation: An overview," *Lang. Linguist. Compass*, vol. 3, no. 2, pp. 537–558, 2009.
4. S. A. Alex, R. Bellad, A. Sumod, and S. P. Sawkar, "A review on scopes and issues in green complier, solving synonym, homonym, hyponym and polysemy problems and translation of English algorithm in C program using SDT," *Int. J. Res. Eng. Sci. Manag.*, vol. 2, no. 5, pp. 316–320, 2019.
5. S. T. Gries, "Polysemy," *Handb. Cogn. Linguist.*, vol. 39, pp. 472–490, 2015.
6. Y. Ravin, "Synonymy from a computational point of view," in A. Lehrer and E. F. Kittay (eds.) *Frames, Fields, and Contrasts*, Routledge, London, 2012, pp. 397–420.

7. J. Morato, M. A. Marzal, J. Lloréns, and J. Moreiro, "Wordnet applications," in *Proceedings of Global Wordnet Conference,* Czech Republic, 2004, pp. 20–23.

8. R. Navigli and F. Martelli, "An overview of word and sense similarity," *Nat. Lang. Eng.*, vol. 25, no. 6, pp. 693–714, 2019.

9. E. Agirre, "Word sense disambiguation: Algorithms and applications," *Springer Google Sch.*, vol. 2, pp. 1166–1174, 2007.

10. R. Kumar, R. Khanna, and V. Goyal, "A review of literature on word sense disambiguation," *Int. J. Eng. Sci.*, vol. 6, pp. 224–230, 2012.

11. A. A. Gadzhiev and A. K. Khmelev, "Lesk algorithm and babelfy system for disambiguation," *Appl. Linguist.*, vol. 36, p. 55.

12. T. Brants, "Natural language processing in information retrieval," *CLIN*, vol. 111, pp. 1–13, 2003.

13. A. F. Smeaton, "Using NLP or NLP resources for information retrieval tasks," in T. Strzalkowski (eds) *Natural Language Information Retrieval,* Springer, Hoboken, NJ, 1999, pp. 99–111.

14. D. D. Lewis and K. S. Jones, "Natural language processing for information retrieval," *Commun. ACM*, vol. 39, no. 1, pp. 92–101, 1996.

4

DISCOURSE AND
PRAGMATIC ANALYSIS

4.1 Important Terms

Discourse refers to a cohesive and structured group of sentences. It goes beyond individual sentences and involves the arrangement of sentences in a way that makes them interconnected and meaningful. Discourse is not just a collection of random sentences; instead, it forms a unified and organized composition. Discourse refers to spoken or written communication, often involving longer stretches of language than individual sentences. It involves the organization of language beyond the sentence level to convey meaning.

Cohesion refers to the grammatical and lexical mechanisms that connect different parts of a text and make it cohesive or stick together. It includes features like pronoun usage, conjunctions, and repetition that link sentences and paragraphs.

Cohesive describes something that is being unified. In the context of language, it refers to the elements within a text being connected and logically related, contributing to the overall flow and clarity.

Discourse structure is the organization and arrangement of elements in spoken or written communication. It involves how sentences and paragraphs are structured to convey a coherent and meaningful message. It encompasses the overall architecture of a piece of discourse.

Adjacency pairs are a concept in conversation analysis, representing pairs of utterances that are closely related and linked. One person's statement or question is followed by a specific and expected response from another person. Examples include question-answer pairs or greeting-response pairs.

DOI: 10.1201/9781003425328-4 **65**

Coherence is used to describe the relationship between sentences that creates a sense of unity and structure within a discourse. Coherence is what distinguishes a well-organized and purposeful arrangement of sentences from a random assortment. It implies that there is a logical and meaningful connection between the sentences, contributing to the overall flow and comprehensibility of the discourse.

Implementation

```
% Python program

import nltk
from nltk.tokenize import sent_tokenize, word_tokenize
from nltk.corpus import stopwords

# Sample text
text = "Discourse refers to spoken or written
communication. Cohesion involves the organization of
language: beyond individual sentences."

# Tokenize sentences and words
sentences = sent_tokenize(text)
words = word_tokenize(text)

# Remove stopwords for cohesion analysis
stop_words = set(stopwords.words('english'))
filtered_words = [word.lower() for word in words if
word.lower() not in stop_words and word.isalpha()]

# Calculate cohesion (example: pronoun usage)
pronoun_count = filtered_words.count('it') + filtered_
words.count('its') + filtered_words.count('they') +
filtered_words.count('them')

# Discourse structure analysis
discourse_structure = {
    'Sentences': sentences,
    'Word_Count': len(words),
    'Cohesive': pronoun_count > 0  # Cohesive if any
pronouns are used
}
```

```
# Display results
print("Sentences:", discourse_structure['Sentences'])
print("Word Count:", discourse_structure['Word_Count'])
print("Cohesive:", discourse_structure['Cohesive'])

% Output:

Sentences: ['Discourse refers to spoken or written
communication.', 'Cohesion involves the organization
of language: beyond individual sentences.']
Word Count: 19
Cohesive: False
```

Implementation of adjacency pairs

```
% Python program

import nltk
from nltk.tokenize import sent_tokenize

# Sample conversation
conversation = "How are you? I'm good, thank you!
What's your name? My name is John. Nice to meet you!
Nice to meet you too!"

# Tokenize into sentences
sentences = sent_tokenize(conversation)

# Identify adjacency pairs (questions and responses)
adjacency_pairs = []

for i in range(0, len(sentences)-1, 2):  # Loop
through sentence pairs
    # Ensure we don't go out of bounds
    pair = (sentences[i], sentences[i+1])
    adjacency_pairs.append(pair)

# Display adjacency pairs
for i, pair in enumerate(adjacency_pairs, 1):
    print(f"Pair {i}:")
    print("Question:", pair[0])
    print("Response:", pair[1])
    print()
```

```
% Output:

Pair 1:
Question: How are you?
Response: I'm good, thank you!

Pair 2:
Question: What's your name?
Response: My name is John.

Pair 3:
Question: Nice to meet you!
Response: Nice to meet you too!
```

4.2 Ethnography of Speaking

Ethnography of speaking allows NLP models to navigate the nuances of language with greater finesse [1]. Ethnography of speaking is an approach that investigates how people use language in their everyday lives within specific cultural and social contexts. It aims to unveil the embedded norms, rituals, and social structures influencing verbal communication. In NLP, incorporating ethnographic insights provides a deeper understanding of language variations and cultural nuances, allowing models to better comprehend and generate contextually appropriate responses. NLP systems, when informed by the ethnography of speaking, can navigate diverse linguistic landscapes more effectively. Cultural sensitivity becomes paramount, enabling models to recognize variations in language use and adapt responses to align with cultural expectations.

Example
User:
   "Hey, what's up?"
NLP System (Trained with Ethnographic Data):
   "Hey! Not much, how can I assist you today?"
   In this example, the NLP system's response reflects an understanding of the informal greeting "Hey, what's up?" based on the ethnographic insights gathered from the community's linguistic practices. This implementation of ethnography of speaking enhances the system's ability to engage users in a culturally sensitive and contextually relevant manner.

4.3 Implicature

Implicature [2] refers to the phenomenon where speakers convey more meaning than explicitly stated. In the context of NLP, recognizing implicatures is crucial for understanding the richness of human communication.

NLP algorithms are designed to identify conversational implicatures by analyzing linguistic nuances and contextual cues. This capability enhances the system's ability to grasp implied meanings and respond appropriately in natural language interactions.

Implicature recognition contributes to more nuanced sentiment analysis in NLP. By discerning subtle implied sentiments, systems can generate responses that align more closely with the underlying emotional tone of the user.

Example
User Inquiry in a Virtual Assistant:

> User: "Is it hot outside?"
> Implicit meaning: The user is likely looking for weather information.

In an NLP system with implicature recognition, the virtual assistant understands the implied intent and responds with the relevant weather forecast without the user explicitly mentioning it.

4.4 Cooperative Principle

The Cooperative Principle, proposed by philosopher H.P. Grice [3], underlines the inherent cooperative nature of communication. It consists of four maxims – quantity, quality, relation, and manner – that speakers adhere to in order to achieve effective communication. NLP models are programmed to adhere to the Cooperative Principle, ensuring that responses generated in dialogue systems align with the principles of informativeness, truthfulness, relevance, and clarity. Understanding the Cooperative Principle aids NLP systems in mitigating potential misunderstandings. By aligning responses with the principles of cooperative communication, systems can generate more contextually relevant and user-friendly outputs.

Example

User: "I can't find my keys."

NLP System: "Did you check the usual spots like your pockets, desk, or the kitchen counter?"

The NLP system adheres to the Cooperative Principle by providing relevant and helpful information, assuming that the user might appreciate suggestions for likely locations where the keys might be found.

Implementation

```
% Python program

import nltk
from nltk.tokenize import word_tokenize
from nltk import pos_tag

# Function to check if a sentence follows the
cooperative principle
def follows_cooperative_principle(sentence):
    # Check if the sentence contains polite and
relevant information
    return "please" in sentence.lower() and
"information" in sentence.lower()

# Function to analyze implicature in a sentence
def analyze_implicature(sentence):
    # Tokenize and tag parts of speech
    tokens = word_tokenize(sentence)
    tagged_tokens = pos_tag(tokens)

    implicature_found = False

    # Check for implicature patterns (simple
pattern-based check)
    for i in range(len(tagged_tokens) - 1):
        # Look for sentences with modal verbs or
negations that might imply something
        if tagged_tokens[i][1] in ['MD', 'VB'] and
tagged_tokens[i+1][1] == 'VB':
            implicature_found = True
            break

    return implicature_found
```

```
# Example conversation
conversation = [
    "Can you please provide me with the information?",
    "I can't do that.",
    "Why not?",
    "I'm currently busy."
]

# Analyze each statement in the conversation
for statement in conversation:
    print("Statement:", statement)

    # Check if the statement follows the cooperative
principle
    if follows_cooperative_principle(statement):
        print("Follows Cooperative Principle: Yes")
    else:
        print("Follows Cooperative Principle: No")

    # Check for implicature in the statement
    if analyze_implicature(statement):
        print("Implicature Found: Yes\n")
    else:
        print("Implicature Found: No\n")

% Output:

Statement: Can you please provide me with the
information?
Follows Cooperative Principle: Yes
Implicature Found: Yes

Statement: I can't do that.
Follows Cooperative Principle: No
Implicature Found: No

Statement: Why not?
Follows Cooperative Principle: No
Implicature Found: No

Statement: I'm currently busy.
Follows Cooperative Principle: No
Implicature Found: No
% Python program
```

```
import nltk
from nltk.tokenize import word_tokenize
from nltk import pos_tag, ne_chunk

# Function to identify entities in a sentence
def identify_entities(sentence):
    tokens = word_tokenize(sentence)  # Tokenize the
sentence
    tagged = pos_tag(tokens)  # Part-of-speech tagging
    entities = ne_chunk(tagged)  # Named entity
recognition
    return entities

# Example usage
sentence = "Apple is planning to launch a new product."
schema_script_result = identify_entities(sentence)

print(schema_script_result)

% Output:

(S
  (GPE Apple/NNP)
  is/VBZ
  planning/VBG
  to/TO
  launch/VB
  a/DT
  new/JJ
  product/NN
  ./.)
```

4.5 Schema-Script

Schema-scripts [4], cognitive structures in the mind, serve as frameworks that help individuals comprehend and predict events. In NLP, these frameworks become essential for machines to understand and interpret the implied structure and order in language. NLP systems utilize schema-script knowledge to understand the expected structure of textual information, enabling them to extract meaningful insights and relationships between different elements in a document. By incorporating schema-scripts, NLP models can predict likely

events or actions based on the context provided, enhancing their ability to generate coherent and contextually relevant responses
Implementation

```
% Python program

import nltk
from nltk.tokenize import sent_tokenize, word_tokenize
from nltk import pos_tag
from nltk.chunk import RegexpParser

# Function to analyze turns and topics in a
conversation
def analyze_turns_and_topics(conversation):
    # Tokenizing the conversation into sentences
    sentences = sent_tokenize(conversation)

    # Analyzing turn-taking patterns
    for i, sentence in enumerate(sentences):
        speaker = "User" if i % 2 == 0 else "Bot"
        print(f"{speaker}: {sentence}")

    # Defining a simple rule-based grammar for topic
shifts
    grammar = """
TOPIC_SHIFT: {<VB.*><.*>*}
    """

    # Chunking sentences and identifying topic shifts
    for sentence in sentences:
        tokens = word_tokenize(sentence)
        tagged = pos_tag(tokens)

        # Create the chunk parser with the defined
grammar
        chunk_parser = RegexpParser(grammar)
        tree = chunk_parser.parse(tagged)

        # Check for subtrees with label 'TOPIC_SHIFT'
        for subtree in tree.subtrees():
            if subtree.label() == 'TOPIC_SHIFT':
                print(f"\nTopic Shift Detected:
{subtree.leaves()}")
```

```
# Example conversation
conversation = """
How are you doing today? I'm good, thanks for asking!
What's the weather like? It's sunny today, perfect for
a walk.
Can you help me with the project? Sure, let's start
with the requirements.
How about the budget? We can discuss the budget once
the project plan is set.
"""

# Analyze the conversation for turns and topic shifts
analyze_turns_and_topics(conversation)

% Output:

User:
How are you doing today?
Bot: I'm good, thanks for asking!
User: What's the weather like?
Bot: It's sunny today, perfect for a walk.
User: Can you help me with the project?
Bot: Sure, let's start with the requirements.
User: How about the budget?
Bot: We can discuss the budget once the project plan
is set.

Topic Shift Detected: [('are', 'VBP'), ('you', 'PRP'),
('doing', 'VBG'), ('today', 'NN'), ('?', '.')]

Topic Shift Detected: [("'m", 'VBP'), ('good', 'JJ'),
(',', ','), ('thanks', 'NNS'), ('for', 'IN'),
('asking', 'VBG'), ('!', '.')]

Topic Shift Detected: [("'s", 'VBZ'), ('the', 'DT'),
('weather', 'NN'), ('like', 'IN'), ('?', '.')]

Topic Shift Detected: [("'s", 'VBZ'), ('sunny', 'JJ'),
('today', 'NN'), (',', ','), ('perfect', 'NN'), ('for',
'IN'), ('a', 'DT'), ('walk', 'NN'), ('.', '.')]

Topic Shift Detected: [('help', 'VB'), ('me', 'PRP'),
('with', 'IN'), ('the', 'DT'), ('project', 'NN'),
('?', '.')]
```

```
Topic Shift Detected: [('let', 'VB'), ("'s", 'POS'),
('start', 'VB'), ('with', 'IN'), ('the', 'DT'),
('requirements', 'NNS'), ('.', '.')]

Topic Shift Detected: [('discuss', 'VB'), ('the',
'DT'), ('budget', 'NN'), ('once', 'IN'), ('the',
'DT'), ('project', 'NN'), ('plan', 'NN'), ('is',
'VBZ'), ('set', 'VBN'), ('.', '.')]
```

4.6 Conversational Analysis

Conversational analysis in NLP [5] is about going beyond the literal meaning of words and understanding the dynamics of human communication. By incorporating these insights, NLP systems can engage in more natural, contextually aware, and meaningful conversations with users. Conversational analysis in NLP, involves examining the structure, patterns, and dynamics of dialogues or conversations to understand how communication unfolds. This analysis is crucial for natural language processing systems to generate more contextually appropriate and coherent responses, as well as to interpret the intended meaning behind user input. Conversational analysis in the context of discourse and pragmatic analysis in NLP is as:

- **Turn-Taking Patterns**: Examining how speakers alternate in a conversation. Helps develop dialogue systems for natural and fluid interactions.
- **Pauses and Interruptions**: Analyzing breaks and disruptions in conversation. Enables systems to respond appropriately to pauses and changes in flow.
- **Conversational Flow**: Assessing the smoothness of conversation. Ensures coherent responses for better user engagement.
- **Shifts in Topic:** Identifying changes in conversation topics. Helps generate contextually relevant responses.
- **Sentiment Analysis**: Analyzing emotional tone in conversation. Recognizes emotions for empathetic and appropriate responses.
- **Pragmatic Implications**: Examining implied meaning in utterances. Aids in understanding indirect speech acts for accurate responses.

Implementation

```
% Python program

# Example conversation 2
conversation_2 = """
How are you today?
I'm doing well, thanks for asking! How about you?
I'm good too, just a bit tired.
I understand, a good rest should help. Do you have any
plans for the day?
I was thinking about going to the park later.
That sounds like a great idea! The weather is perfect
for it.
Yeah, it's been sunny all day.
Hopefully, it stays nice for you. What will you do in
the park?
"""

# Analyze the conversation for turns and topic shifts
(Conversation 2)
print("Conversation 2 Analysis:\n")
analyze_turns_and_topics(conversation_2)

% Output:

Conversation 2 Analysis:

User:
How are you today?
Bot: I'm doing well, thanks for asking!
User: How about you?
Bot: I'm good too, just a bit tired.
User: I understand, a good rest should help.
Bot: Do you have any plans for the day?
User: I was thinking about going to the park later.
Bot: That sounds like a great idea!
User: The weather is perfect for it.
Bot: Yeah, it's been sunny all day.
User: Hopefully, it stays nice for you.
Bot: What will you do in the park?

Topic Shift Detected: [('are', 'VBP'), ('you', 'PRP'),
('today', 'NN'), ('?', '.')]
```

```
Topic Shift Detected: [("'m", 'VBP'), ('doing',
'VBG'), ('well', 'RB'), (',', ','), ('thanks', 'NNS'),
('for', 'IN'), ('asking', 'VBG'), ('!', '.')]

Topic Shift Detected: [("'m", 'VBP'), ('good', 'JJ'),
('too', 'RB'), (',', ','), ('just', 'RB'), ('a',
'DT'), ('bit', 'NN'), ('tired', 'JJ'), ('.', '.')]

Topic Shift Detected: [('understand', 'VBP'), (',',
','), ('a', 'DT'), ('good', 'JJ'), ('rest', 'NN'),
('should', 'MD'), ('help', 'VB'), ('.', '.')]

Topic Shift Detected: [('Do', 'VBP'), ('you', 'PRP'),
('have', 'VB'), ('any', 'DT'), ('plans', 'NNS'), ('for',
'IN'), ('the', 'DT'), ('day', 'NN'), ('?', '.')]

Topic Shift Detected: [('was', 'VBD'), ('thinking',
'VBG'), ('about', 'RB'), ('going', 'VBG'), ('to',
'TO'), ('the', 'DT'), ('park', 'NN'), ('later', 'RB'),
('.', '.')]

Topic Shift Detected: [('sounds', 'VBZ'), ('like',
'IN'), ('a', 'DT'), ('great', 'JJ'), ('idea', 'NN'),
('!', '.')]

Topic Shift Detected: [('is', 'VBZ'), ('perfect',
'JJ'), ('for', 'IN'), ('it', 'PRP'), ('.', '.')]

Topic Shift Detected: [(''', 'VBD'), ('s', 'RB'),
('been', 'VBN'), ('sunny', 'JJ'), ('all', 'DT'),
('day', 'NN'), ('.', '.')]

Topic Shift Detected: [('stays', 'VBZ'), ('nice',
'JJ'), ('for', 'IN'), ('you', 'PRP'), ('.', '.')]

Topic Shift Detected: [('do', 'VB'), ('in', 'IN'),
('the', 'DT'), ('park', 'NN'), ('?', '.')]
```

4.7 Deciphering Meaning and Coherence of Text Data

Concepts like endophora, exophora, and various types of context
play a crucial role in deciphering the meaning and coherence of textual data.

4.7.1 Endophora

Endophora [6] refers to the reference within a text, where a word or phrase refers to something previously mentioned or alluded to in the same discourse. Recognizing and resolving endophoric references is essential for maintaining coherence in a document or conversation.

Example

User: I bought a new phone. It has an amazing camera.

In this example, "It" in the second sentence is an example of endophora, referring back to the previously mentioned "new phone."

4.7.2 Exophora

Exophora [7], on the other hand, involves references outside the current text or discourse. The reference is established beyond the immediate linguistic context. Understanding exophoric references is crucial for interpreting the intended meaning when the context lies outside the current text.

Example

User: Look at that beautiful building!

In this case, "that" is an exophoric reference as it points to something external to the text, and the interpretation depends on the situational context.

Implementation

```
% Python program

import nltk
from nltk.tokenize import word_tokenize
from nltk.tag import pos_tag
from nltk.chunk import ne_chunk

# Function to identify endophora and exophora in a
sentence
def identify_endophora_exophora(sentence):
    # Tokenize and tag the sentence
    tokens = word_tokenize(sentence)
    tagged_tokens = pos_tag(tokens)
```

```
    # Perform Named Entity Recognition (NER)
    named_entities = ne_chunk(tagged_tokens)

    # Identify and print named entities in the
sentence
    named_entity_found = False
    for subtree in named_entities:
        if isinstance(subtree, nltk.Tree):  # A
subtree is a named entity
            entity_label = subtree.label()
            # Look for common named entity labels like
'PERSON', 'GPE' (Geopolitical Entity), etc.
            if entity_label in ['PERSON', 'GPE',
'ORGANIZATION']:
                named_entity_found = True
                print(f"Named entity: {subtree.
leaves()} ({entity_label})")

    if not named_entity_found:
        print("No named entities found in the
sentence.")

    # Identify pronouns and check for endophora or
exophora
    pronouns = [token for token, pos in tagged_tokens
if pos == 'PRP']

    for pronoun in pronouns:
        if pronoun.lower() in ['he', 'she', 'it']:
            # Exophoric expression referring outside
the text
            print(f"Exophoric expression: {pronoun}")
        else:
            # Endophoric expression referring within
the text
            print(f"Endophoric expression: {pronoun}")

# Example usage
sentence = "John visited the museum. He enjoyed the
exhibits, and it was a great experience."
identify_endophora_exophora(sentence)
```

```
% Output:

Named entity: [('John', 'NNP')] (PERSON)
Exophoric expression: He
Exophoric expression: it
```

4.8 Discourse Context and Its Types

Discourse context in NLP [8] refers to the broader context within which individual utterances or sentences are situated. It encompasses the surrounding linguistic environment, such as preceding and following sentences, paragraphs, or even entire conversations or documents. Understanding discourse context is crucial for interpreting the intended meaning of a specific expression and is an essential aspect of advanced language understanding in NLP.

Importance of Discourse Context:

- Discourse context is crucial for resolving coreferences, such as pronouns, by identifying the entities they refer to within the discourse.
- NLP systems generating text benefit from understanding and maintaining coherence with the existing discourse context, ensuring that responses are contextually relevant and flow naturally.
- Recognizing the discourse context helps NLP models grasp the nuances, intentions, and relationships between different elements in a text, leading to more accurate language comprehension.

Types of Discourse Context:

- **Anaphoric Context**

   Anaphoric context involves the relationship between a word or phrase and its antecedent (a word or phrase mentioned earlier in the text). Resolving anaphoric references is crucial for understanding which entities are being referred to within the discourse.

   Example: "John bought a new car. He loves the features."

   In this example, "He" is an anaphoric reference that points back to "John" in the previous sentence.

- **Cataphoric Context**

    Cataphoric context, in contrast to anaphoric context, occurs when a word or phrase refers to something mentioned later in the discourse. This requires anticipation of information that follows in the text.

    Example: "After the long day, John arrived home. There, he found a surprise waiting for him."

    Here, "There" is a cataphoric reference that anticipates the information about the surprise, which follows in the subsequent sentence.

- **Exophoric Context**

    Exophoric context involves references to elements outside the text, relying on shared knowledge between the speaker and the listener. This can include references to objects, locations, or events in the real-world context.

    Example: "Look at that beautiful sunset!"

    The term "that" in this context relies on shared visual or situational knowledge between the speaker and the listener.

Implementation

```
% Python program

import nltk
from nltk.tokenize import sent_tokenize, word_tokenize
from nltk.tag import pos_tag

def analyze_discourse_context(text):
    sentences = sent_tokenize(text)  # Tokenize the
text into sentences

    for i in range(len(sentences) - 1):
        current_sentence = sentences[i]
        next_sentence = sentences[i + 1]

        current_tokens = word_tokenize(current_
sentence)  # Tokenize current sentence
        next_tokens = word_tokenize(next_sentence)  #
Tokenize next sentence
```

```
    current_pos_tags = pos_tag(current_tokens)   #
Part-of-speech tagging for current sentence

    # Anaphoric context (refers to something
mentioned earlier in the text)
    anaphoric_references = [token for token, pos
in current_pos_tags if pos == 'PRP' and token.lower()
in ['he', 'she', 'it']]

    if anaphoric_references:
        print(f"Anaphoric reference in '{current_
sentence}': {anaphoric_references} refers to something
in the previous sentence.")

    # Cataphoric context (anticipates something in
the next sentence)
    cataphoric_references = [next_tokens[0] for
token, pos in current_pos_tags if pos == 'DT' and
token.lower() in ['this', 'that']]

    if cataphoric_references:
        print(f"Cataphoric reference in '{current_
sentence}': {cataphoric_references} anticipates
something in the next sentence.")

    # Exophoric context (relies on shared
knowledge outside the text)
    exophoric_references = [token for token in
current_tokens if token.lower() in ['there', 'that']]

    if exophoric_references:
        print(f"Exophoric reference in '{current_
sentence}': {exophoric_references} relies on shared
knowledge outside the text.")

# Example usage
text = "John bought a new car. He loves the features.
This car is impressive. Look at that beautiful sunset!"
analyze_discourse_context(text)

% Output:

Anaphoric reference in 'He loves the features.':
['He'] refers to something in the previous sentence.

Cataphoric reference in 'This car is impressive.':
['Look'] anticipates something in the next sentence.
```

4.9 Speech Acts

In pragmatic analysis, speech acts [9] refer to the actions that speakers perform with their utterances beyond conveying literal meanings. These actions can be categorized into two main types: direct speech acts and indirect speech acts. Understanding these distinctions is essential in NLP for machines to accurately interpret and respond to user inputs. In NLP, recognizing and understanding speech acts, whether direct or indirect, is crucial for machines to generate appropriate responses. This involves not only parsing the literal meaning of the words but also grasping the intended illocutionary force behind the utterance. NLP systems that incorporate pragmatic analysis can better handle user requests, commands, and queries by considering the broader context and implied meanings in communication. This leads to more contextually relevant and human-like interactions between users and machines.

4.9.1 Direct Speech Act

Direct speech acts involve explicit and straightforward communication, where the speaker's intended meaning aligns with the literal interpretation of the words used. In other words, the speaker directly conveys their intention through the uttered sentence without relying on additional context or implicit cues.

Example

> Request: "Pass me the salt."
> Command: "Close the door."
> Statement: "I will be there at 3 PM."
> Question: "What is your name?"

In each of these examples, the speaker's intention is clear, and the meaning of the utterance is directly correlated with the illocutionary force (intended action) of the speech act.

4.9.2 Indirect Speech Act

Indirect speech acts [10] involve a more nuanced form of communication, where the speaker conveys their intention indirectly, often relying on context, shared knowledge, or pragmatic inference. The literal meaning of the words may not directly correspond to the intended illocutionary force.

Example

Request (Indirect): "It's chilly in here." (intended illocutionary force: "Close the window.")

Command (Indirect): "I wonder if you could pass me the salt." (intended illocutionary force: "Pass me the salt.")

Statement (Indirect): "The trash is overflowing." (intended illocutionary force: "Take out the trash.")

Question (Indirect): "Do you have the time?" (intended illocutionary force: "Tell me the time.")

In these examples, the speaker's intention is conveyed indirectly, and the listener is expected to infer the illocutionary force based on the context and pragmatic cues

Implementation

```
% Python program

def classify_speech_act(utterance):
    # Direct speech acts
    direct_requests = ["please", "could you", "would
you"]
    direct_commands = ["shut", "close", "open"]
    direct_statements = ["I will", "It is", "The sky is"]
    direct_questions = ["what", "where", "when",
"who", "how"]

    # Check for direct speech acts
    if any(word in utterance.lower() for word in
direct_requests):
        print("Direct Request Detected")

    elif any(word in utterance.lower() for word in
direct_commands):
        print("Direct Command Detected")

    elif any(word in utterance.lower() for word in
direct_statements):
        print("Direct Statement Detected")
```

```
    elif any(word in utterance.lower() for word in
direct_questions):
        print("Direct Question Detected")

    else:
        # Check for indirect speech acts
        if "wonder" in utterance.lower():
            print("Indirect Request Detected")

        elif "if" in utterance.lower() and "could you"
in utterance.lower():
            print("Indirect Command Detected")

        elif "it's" in utterance.lower() or "isn't" in
utterance.lower():
            print("Indirect Statement Detected")

        elif "wonder" in utterance.lower():
            print("Indirect Question Detected")

        else:
            print("Speech Act Not Detected")

# Example usage:

classify_speech_act("Please close the door.")
classify_speech_act("I wonder if you could pass the
salt.")
classify_speech_act("What is your name?")
classify_speech_act("The dishes aren't washing
themselves.")

% Output:

Direct Request Detected
Indirect Request Detected
Direct Question Detected
Speech Act Not Detected
```

4.10 Deixis and Deictic Expressions

Deixis [11] refers to words and phrases that show time, place, or situation when someone is talking. It refers to words or phrases such as "me," "here," etc., which are difficult to understand without additional information.Deixis is a linguistic phenomenon that involves the use of words or expressions whose interpretation relies heavily on the context of the conversation, particularly the time, place, or situation in which the communication is taking place. These words or phrases, known as deictic expressions, contribute to the understanding of the message by referring to elements that are not explicitly stated but are instead inferred from the surrounding context. Deictic expressions, which convey meaning based on context, can be classified into three main types:

- **Person Deixis:** This refers to expressions that indicate the participants in the communication, such as pronouns. Relates to pronouns (e.g., "I," "you") whose meaning depends on the participants in the conversation.
- **Spatial Deixis**: Spatial deictic expressions are those that convey information about the location of entities in space. Involves words like "here" and "there" that convey location, relying on the physical context of the speaker and listener.
- **Temporal Deixis:** Temporal deictic expressions provide information about the timing of events or actions. Refers to words such as "now" and "tomorrow" that convey timing, with meaning dependent on when the communication occurs.

Implementation

```
% Python program

import nltk
from nltk.tokenize import word_tokenize
from nltk.tag import pos_tag

# Function to classify deictic expressions
def classify_deictic_expression(sentence):
    tokens = word_tokenize(sentence)
    tagged_tokens = pos_tag(tokens)

    for token, pos in tagged_tokens:
        if pos == "PRP":  # Person deixis (pronouns)
            print(f"{token} is a person deictic
expression.")
```

```
        elif pos == "RB" and token.lower() in ['here',
'there']:  # Spatial deixis (adverbs indicating
location)
            print(f"{token} is a spatial deictic
expression.")

        elif pos == "NN" and token.lower() in ['today',
'tomorrow', 'yesterday']:  # Temporal deixis (nouns
indicating time)
            print(f"{token} is a temporal deictic
expression.")

# Example usage:
sentence1 = "Meet me here."
sentence2 = "I wish you'd been here yesterday."
sentence3 = "I'll see you tomorrow."

print("Sentence 1:")
classify_deictic_expression(sentence1)

print("\nSentence 2:")
classify_deictic_expression(sentence2)

print("\nSentence 3:")
classify_deictic_expression(sentence3)

% Output:

Sentence 1:
me is a person deictic expression.
here is a spatial deictic expression.

Sentence 2:
I is a person deictic expression.
you is a person deictic expression.
here is a spatial deictic expression.
yesterday is a temporal deictic expression.

Sentence 3:
I is a person deictic expression.
you is a person deictic expression.
tomorrow is a temporal deictic expression.
% Python program
```

```
from nltk.sentiment import SentimentIntensityAnalyzer

def analyze_positive_negative_face(sentence):
    sia = SentimentIntensityAnalyzer()
    sentiment_scores = sia.polarity_scores(sentence)

    # Positive Face
    if sentiment_scores['compound'] >= 0.05:
        print("The statement conveys a positive
sentiment.")

    # Negative Face
    elif sentiment_scores['compound'] <= -0.05:
        print("The statement conveys a negative
sentiment.")

    # Neutral Face
    else:
        print("The sentiment of the statement is
neutral.")

# Example Usage
user_input = "I appreciate your help with the
project."
analyze_positive_negative_face(user_input)

% Output:

The statement conveys a positive sentiment.
```

4.11 Positive and Negative Face in Pragmatics

In pragmatic analysis [12], positive face and negative face are concepts introduced by sociolinguist Erving Goffman and later developed by politeness theorists, particularly Brown and Levinson. These concepts are crucial in understanding how individuals manage their social interactions through language. Positive face refers to the desire for approval, liking, and a sense of being valued, while negative face pertains to the desire for autonomy, freedom from imposition, and the right to act without interference.

4.11.1 Positive Face

Positive face represents an individual's need to be appreciated, liked, and affirmed in their social interactions. It involves the desire for a sense of belonging and positive regard from others. In natural language processing (NLP), understanding positive face is crucial for developing socially intelligent systems that can generate responses that promote a positive and supportive interaction with users. Politeness strategies, such as using courteous language or expressing appreciation, can contribute to a more positive user experience.

4.11.2 Negative Face

Negative face reflects an individual's need for autonomy, independence, and the freedom to act without being imposed upon. It involves the desire to have one's actions and decisions respected and not encroached upon by others. Recognizing negative face in NLP is essential for developing systems that respect users' autonomy and preferences. Language models can be designed to offer choices, seek permission, and use polite language to minimize the potential imposition on users.

Implementation

```
% Python program

from nltk.sentiment import SentimentIntensityAnalyzer

def analyze_positive_negative_face(sentence):
    sia = SentimentIntensityAnalyzer()
    sentiment_scores = sia.polarity_scores(sentence)

    # Positive Face
    if sentiment_scores['compound'] >= 0.05:
        print("The statement conveys a positive
sentiment.")

    # Negative Face
    elif sentiment_scores['compound'] <= -0.05:
```

```
        print("The statement conveys a negative
sentiment.")

    # Neutral Face
    else:
        print("The sentiment of the statement is
neutral.")

# Example Usage
user_input = "I appreciate your help with the
project."
analyze_positive_negative_face(user_input)

% Output:

The statement conveys a positive sentiment.
% Python program

from nltk.sentiment import SentimentIntensityAnalyzer

def analyze_politeness(text):
    sia = SentimentIntensityAnalyzer()
    sentiment_scores = sia.polarity_scores(text)

    # Positive face: Higher positive sentiment
indicates a positive response
    positive_face = sentiment_scores['pos']

    # Negative face: Higher negative sentiment
indicates a negative response
    negative_face = sentiment_scores['neg']
    return positive_face, negative_face

# Example usage
user_input = "Thank you for your assistance, but I
have some concerns about the product."

positive, negative = analyze_politeness(user_input)

print(f"Positive Face Score: {positive}")
print(f"Negative Face Score: {negative}")
```

```
% Output:

Positive Face Score: 0.137
Negative Face Score: 0.0
% Python program

import random
import time

def chatbot_response(user_input):
    # Greeting markers
    greeting_markers = ["Hello!", "Hi there!", "Hey!"]
    # Politeness markers
    politeness_markers = ["Certainly,", "Please,", "I
would be happy to,"]
    # Emphasis markers
    emphasis_markers = ["Absolutely!", "Indeed,",
"Certainly!"]

    # Randomly select a greeting marker
    greeting_marker = random.choice(greeting_markers)
    # Randomly select a politeness marker
    politeness_marker = random.choice(politeness_markers)
    # Randomly select an emphasis marker
    emphasis_marker = random.choice(emphasis_markers)

    # Check for specific input types and respond
accordingly
    user_input_lower = user_input.lower()

    # Handle greetings
    if "hello" in user_input_lower or "hi" in
user_input_lower:
        return f"{greeting_marker} How can I assist
you today?"

    # Handle polite requests
    if "please" in user_input_lower or "kindly" in
user_input_lower:
        return f"{politeness_marker} Thank you for
asking. How may I help you?"
```

```
    # Handle urgency and important requests
    if "urgent" in user_input_lower or "important" in
user_input_lower:
        return f"{emphasis_marker} I understand this
is urgent. Please provide more details for quicker
assistance."

    # Handle questions about time
    if "time" in user_input_lower:
        current_time = time.strftime("%I:%M %p")
        return f"The current time is {current_time}.
How else can I assist you?"

    # Handle general questions
    if "how are you" in user_input_lower or "how's it
going" in user_input_lower:
        return f"I'm doing great, thank you for asking!
How can I assist you today?"

    # Handle questions about the chatbot itself
    if "who are you" in user_input_lower or "what is
your name" in user_input_lower:
        return "I am your friendly assistant, here to
help you with anything you need!"

    # Handle random statements or fallback responses
    return "I'm not sure how to respond. Could you
please provide more context or ask a specific question?"

# Example usage
while True:
    user_input = input("User: ")
    response = chatbot_response(user_input)
    print("Chatbot:", response)
    # Optionally, exit the loop if the user wants to
quit the conversation
    if user_input.lower() in ['exit', 'quit', 'bye']:
        print("Chatbot: Goodbye! Have a great day!")
        break

% Output:

User: Hi
Chatbot: Hey! How can I assist you today?
User: What is the date day after tomorrow?
```

```
Chatbot: I'm not sure how to respond. Could you please
provide more context or ask a specific question?
User: bye
Chatbot: I'm not sure how to respond. Could you please
provide more context or ask a specific question?
Chatbot: Goodbye! Have a great day!
```

4.12 Pragmatic Markers and Functions

Pragmatic markers [13], as linguistic cues, convey information about a speaker's attitude, intention, or discourse context. They are vital for nuanced communication, guiding listeners to interpret meaning beyond literal words. In pragmatic analysis, understanding these markers is crucial. In NLP, incorporating pragmatic markers is essential for machines to generate contextually appropriate responses, recognize speaker intentions, and engage in natural conversations. This is particularly valuable in NLP applications like chatbots, enhancing the overall quality of human-machine interaction.

4.12.1 Functions of Pragmatic Markers

- Pragmatic markers serve to highlight and give importance to specific details or expressions within a sentence. They play a role in enhancing the politeness of language, aiding systems in generating responses that are courteous and respectful.
- Pragmatic markers convey the speaker's stance or assessment of a situation, offering a glimpse into their perspective.
- They assist in arranging information in a sequential manner, facilitating the organization of discourse.
- Pragmatic markers contribute to steering the conversation, indicating when a speaker takes a turn or signaling pauses.
- They introduce contrasting ideas or acknowledge opposing viewpoints, recognizing counterarguments.
- Pragmatic markers offer additional explanations or clarifications to ensure better comprehension.
- Hedging: They express uncertainty or moderate statements to convey a less assertive tone.

Implementation

```python
% Python program

import random
import time

# User's data
user_data = {}

# Response markers
greeting_markers = ["Hello!", "Hi there!", "Hey!"]
politeness_markers = ["Certainly,", "Please,", "I
would be happy to,"]
emphasis_markers = ["Absolutely!", "Indeed,",
"Certainly!"]

# Function to handle user name
def ask_for_name():
    return "I don't think we've met. What is your name?"

# Function to greet the user
def greet_user(name):
    return f"Hello, {name}! How can I assist you today?"

# Enhanced chatbot response function
def chatbot_response(user_input):
    global user_data
    greeting_marker = random.choice(greeting_markers)
    politeness_marker = random.choice(politeness_markers)
    emphasis_marker = random.choice(emphasis_markers)

    user_input_lower = user_input.lower()

    # Handle Name Input
    if 'name' in user_input_lower and user_data.
get("name") is None:
        user_name = input("Please enter your name: ")
        user_data["name"] = user_name
        return greet_user(user_name)

    if user_data.get("name"):
        # Use the name in conversations
```

```
        if "hello" in user_input_lower or "hi" in
user_input_lower:
            return greet_user(user_data["name"])

    # Handle greeting
    if "hello" in user_input_lower or "hi" in
user_input_lower:
        return f"{greeting_marker} How can I assist
you today?"

    # Handle polite requests
    if "please" in user_input_lower or "kindly" in
user_input_lower:
        return f"{politeness_marker} Thank you for
asking. How may I help you?"

    # Handle urgency or emphasis
    if "urgent" in user_input_lower or "important" in
user_input_lower:
        return f"{emphasis_marker} I understand this
is urgent. Please provide more details for quicker
assistance."

    # Handle general questions
    if "how are you" in user_input_lower or "how's it
going" in user_input_lower:
        return f"I'm doing great, thank you for asking!
How can I assist you today?"

    # Handle questions about the chatbot
    if "who are you" in user_input_lower or "what is
your name" in user_input_lower:
        return "I am your friendly assistant, here to
help you with anything you need!"

    # Handle current time request
    if "time" in user_input_lower:
        current_time = time.strftime("%I:%M %p")
        return f"The current time is {current_time}.
How else can I assist you?"

    # Handle responses for "ok" or "bye"
    if "ok" in user_input_lower:
```

```
        return "Is there anything else you'd like
assistance with?"

    if "bye" in user_input_lower or "quit" in user_
input_lower or "exit" in user_input_lower:
        return "Goodbye! Have a great day!"

    # Handle vague responses or requests
    if "help" in user_input_lower or "assist" in
user_input_lower:
        return "How can I assist you? Could you please
clarify your request?"

    # Random small talk
    if "joke" in user_input_lower or "tell me a joke"
in user_input_lower:
        return "Why don't skeletons fight each other?
They don't have the guts!"

    # Fallback response for unclear inputs
    return "I'm not sure how to respond. Could you
please provide more context or ask a specific question?"

# Example usage
while True:
    user_input = input("User: ")
    response = chatbot_response(user_input)
    print("Chatbot:", response)

    # Optionally, exit the loop if the user wants to quit
    if user_input.lower() in ['exit', 'quit', 'bye']:
        break

% Output:

User: Hi
Chatbot: Hey! How can I assist you today?
User: who are you
Chatbot: I am your friendly assistant, here to help
you with anything you need!
User: how are you
Chatbot: I'm doing great, thank you for asking! How can
I assist you today?
User: what is the time now
Chatbot: The current time is 07:01 PM. How else can I
assist you?
```

```
User: ok
Chatbot: Is there anything else you'd like assistance
with?
User: bye
Chatbot: Goodbye! Have a great day!
```

4.13 Summary

This chapter explores discourse and pragmatic analysis, which go beyond individual words and sentences to understand how language functions in communication. It begins with the concept of discourse, focusing on cohesion and cohesive devices, which ensure logical connections between sentences. The structure of discourse is examined through elements like adjacency pairs (turn-taking in conversations) and the ethnography of speaking, which studies cultural and social influences on communication. This chapter also covers implicature and the cooperative principle, which explain how speakers convey meaning beyond literal words. Concepts like schema and script theory provide insights into how prior knowledge influences language interpretation. Conversational analysis is discussed to highlight patterns in dialogues, while context and its types (endophora and exophora) are explored to show how meaning depends on situational references.

A key focus is on speech acts, distinguishing between direct and indirect speech acts, and the role of deixis and deictic expressions in pointing to time, place, and people. This chapter concludes with an analysis of pragmatic markers, which help in structuring conversations, and the concepts of positive and negative face, which relate to politeness and social interaction in communication. This chapter provides essential insights into how meaning is constructed beyond individual words, making it a crucial aspect of NLP and linguistic studies.

References

1. T. Arnold and H. J. A. Fuller, "In search of the user's language: Natural language processing, computational ethnography, and error-tolerant interface design," in Advances in usability, user experience and assistive technology: Proceedings of the AHFE 2018 international conferences on usability & user experience and human factors and assistive technology, held on July 21--25, 2018, in Loews Sapphire Falls Resort at, 2019, pp. 36–43.

2. J. C. Sedivy, "Implicature during real time conversation: A view from language processing research," *Philos. compass*, vol. 2, no. 3, pp. 475–496, 2007.

3. J. Thomas, "Cooperative principle," Concise Encycl. Philos. Lang. Peter V. Lamarque, 1997 Philos. Lang., vol. 1, p. 393, 1997.

4. H. I. Joseph, "Narrative Schema as World Knowledge for Coreference Resolution," 2012.

5. Upreti, "A Comparative Analysis of NLP Algorithms For Implementing AI Conversational Assistants: Comparative Analysis of NLP Algorithms for NLI." 2023.

6. M. Safdar, M. J. I. Khan, and P. Hussain, "A review different approach for anaphora resolution".

7. X. Yu, H. Zhang, Y. Song, C. Zhang, K. Xu, and D. Yu, "Exophoric Pronoun Resolution in Dialogues with Topic Regularization," *arXiv Prepr.* arXiv2109.04787, 2021.

8. J. Hirschberg and C. D. Manning, "Advances in natural language processing," *Science (80-.).* vol. 349, no. 6245, pp. 261–266, 2015.

9. C. R. Perrault and J. Allen, "Speech acts as a basis for understanding dialogue coherence," in Theoretical issues in natural language processing-2, 1978.

10. N. Asher and A. Lascarides, "Indirect speech acts," *Synthese*, vol. 128, pp. 183–228, 2001.

11. Dylgjeri and L. Kazazi, "Deixis in modern linguistics and outside," *Acad. J. Interdiscip. Stud.*, vol. 2, no. 4, pp. 87–96, 2013.

12. H. Jucker, "Positive and negative face as descriptive categories in the history of English," *J. Hist. Pragmat.*, vol. 12, no. 1–2, pp. 178–197, 2011.

13. K. Aijmer and A.-M. Simon-Vandenbergen, "Pragmatic markers," *Discursive Pragmat.*, vol. 8, pp. 223–247, 2011.

5

ARTIFICIAL INTELLIGENCE IN NLP

In NLP, artificial intelligence (AI) refers to the application of sophisticated computational methods that allow machines to comprehend, interpret, and produce human language. NLP is a branch of AI that focuses on the way computers interact with natural language, including text and speech.

5.1 Machine Learning

Within the field of AI, machine learning focuses on creating models and algorithms that let computers learn and make judgments or predictions without explicit programming. Algorithms are trained using data so they can recognize patterns, forecast outcomes, and gradually get better at what they do. Numerous fields, including image identification, NLP, recommendation systems, and even more, use machine learning to improve computer system performance and automate operations.

5.1.1 Supervised Machine Learning

Using a labeled dataset, supervised learning entails training a model with input data (features) combined with matching labels or intended outputs. Labeled data in NLP could include phrases or documents that have sentiment tags, named entities, categories, or other predetermined annotations. Labeled datasets, in which the data being input (text) is linked to matching labels or categories, are used to train AI models. This method is frequently applied to applications such as text categorization, named entity recognition, and sentiment analysis.

DOI: 10.1201/9781003425328-5 **99**

Applications in NLP:

- **Sentiment Analysis** [1]: Using labeled examples to categorize text as neutral, negative, or positive.
- **Named Entity Recognition (NER)** [2]: Recognizing named entities, places, or organizations within a text.
- **Text Classification** [3]: Categorizing documents into predefined classes or topics.

5.1.2 Unsupervised Machine Learning

When a model is trained on an unlabeled dataset, unsupervised learning occurs when the algorithm finds patterns, structures, or correlations in the data without direct supervision. Algorithms identify structures and patterns in data without the need for labeling. Word embeddings, topic modeling, and clustering are a few instances of unsupervised learning used in NLP.

Applications in NLP:

- **Topic Modeling** [4]: Finding underlying themes in a set of papers without established classifications.
- **Clustering** [5]: Grouping similar documents or sentences together based on their content.
- **Word Embedding** [6]: Creating dense vector representations of words based on contextual information.

5.2 Machine Learning on Natural Language Sentences

Enabling machines to comprehend, analyze, and produce text that is similar to that of a person requires machine learning on natural language phrases. This helps with applications that range from sentiment assessment and language translation to chatbots and virtual assistants. Utilizing models and algorithms to process and comprehend human language is known as machine learning on natural language sentences. The models are able to generalize and predict new, unseen phrases because they have learned patterns and relationships in the data. The following tasks apply machine learning to natural language sentences:

1. **Text Classification:** Giving text labels or predetermined categories.
2. **Named Entity Recognition:** Recognizing entities in a text, including names, places, or organizations.
3. **Part-of-Speech Tagging:** Giving each word in a sentence a grammatical label (such as noun, verb, or adjective).
4. **Machine Translation:** Converting text between languages.
5. **Text Summarization** [7]: Producing succinct and logical synopses of lengthy texts.
6. **Question Answering** [8]: Comprehending inquiries and offering pertinent responses.
7. **Intent Recognition:** Determining the intention or goal of user input.
8. **Text Generation:** The goal is to use learnt patterns to produce text that appears human.
9. **Word Embeddings:** Using dense vectors to represent words in a continuous vector space is known as word embedding.
10. **Sentiment Analysis** [9]: identifying if a statement expresses a good, negative, or neutral sentiment).

Implementation

```
% Python program

import nltk

from nltk.sentiment import SentimentIntensityAnalyzer

def analyze_sentiment(sentence):
    # Initialize SentimentIntensityAnalyzer
    sia = SentimentIntensityAnalyzer()

    # Get sentiment scores
    sentiment_scores = sia.polarity_scores(sentence)

    # Determine sentiment based on the compound score
    if sentiment_scores['compound'] >= 0.05:
        return 'Positive'
    elif sentiment_scores['compound'] < -0.05:
        return 'Negative'
```

```
    else:
        return 'Neutral'

# Example sentences for sentiment analysis
sentences = [
    "I love this product! It's amazing.",
    "The service was terrible, and I'm very
disappointed.",
    "The weather is nice today.",
    "The movie was neither good nor bad."
]

# Analyze sentiment for each sentence
for sentence in sentences:
    sentiment = analyze_sentiment(sentence)
    print(f"Sentence: {sentence}\nSentiment:
{sentiment}\n")

% Output:

Sentence: I love this product! It's amazing.
Sentiment: Positive

Sentence: The service was terrible, and I'm very
disappointed.
Sentiment: Negative

Sentence: The weather is nice today.
Sentiment: Positive

Sentence: The movie was neither good nor bad.
Sentiment: Negative
```

5.3 Hybrid Machine Learning Systems in NLP

In NLP, hybrid machine learning systems are those that combine several machine learning methods or approaches to capitalize on their unique advantages and get around their drawbacks. In hybrid systems, supervised and unsupervised learning components are combined, or rule-based systems are combined with statistical or deep learning models. The objective is to improve the resilience, flexibility, and performance of NLP applications. Typical forms of hybrid NLP machine learning systems:

Transfer Learning: Using a smaller, task-specific dataset to refine a model for a particular NLP task after it has been trained on a larger, general-purpose dataset is known as transfer learning.

Application: Large volumes of text data are used to train pre-trained models including BERT and GPT, which can then be optimized for tasks like named entity recognition, sentiment analysis, and question answering.

Rule-Based Systems Combined with ML Models: To capture intricate linguistic patterns and relationships, hybrid systems may combine machine learning models with rule-based systems that employ pre-established linguistic rules.

Application: Combining rule-based named entity recognition with machine learning models for more accurate and context-aware entity recognition.

Semi-Supervised Learning: For training, semi-supervised learning combines labeled and unlabeled data. A greater pool of unlabeled data is added to a smaller amount of labeled data.

Application: This approach is useful when labeled data is scarce. For instance, combining a greater volume of unlabeled data with a small classified sample for sentiment analysis.

Ensemble Learning: Compared to individual models, ensemble approaches integrate several machine learning models to get a forecast that is more reliable and accurate.

Application: Developing a collection of several models, such as integrating neural networks, decision trees, and support vector machines, to enhance overall performance in tasks like sentiment analysis or text classification.

Hybrid Neural Networks: To capitalize on their complementing qualities, various neural network design types are combined into a single model.

Application: Combining Recurrent Neural Networks (RNNs) and Convolutional Neural Networks (CNNs) to identify local and sequential patterns in information in text.

Symbolic and Subsymbolic Integration: Integrating symbolic reasoning with subsymbolic machine learning methods to benefit from the interpretability of rule-based systems and the learning capabilities of statistical models.

Application: Combining rule-based syntactic parsing with machine learning models for enhanced parsing accuracy and linguistic understanding.

Implementation

Algorithm

1. **Input:** Take an input text for analysis.
2. Rule-Based Named Entity Recognition:
   a. Tokenize the input text into sentences.
   b. Tokenize each sentence into words.
   c. Part-of-speech tag each word in the sentences.
   d. Apply named entity recognition using the NLTK's ne_ chunk function on the part-of-speech tagged sentences.
   e. Extract named entities from the named entity recognition results.
3. Machine Learning Sentiment Analysis:
   a. Use the NLTK's Sentiment Intensity Analyzer to analyze the sentiment of the input text.
   b. Determine the sentiment type using the compound score: positive, negative, or neutral.
13. Output:
   Display the identified named entities.
   Display the determined sentiment of the input text

```
% Python program

import nltk
from nltk.tokenize import sent_tokenize, word_tokenize
from nltk.tag import pos_tag
from nltk.chunk import ne_chunk
from nltk.sentiment import SentimentIntensityAnalyzer
from nltk.corpus import stopwords

def rule_based_ner(text):
    # Sentence tokenization
    sentences = sent_tokenize(text)

    # Tokenization of each sentence
    tokenized_sentences = [word_tokenize(sentence) for
sentence in sentences]
```

```python
    # Part-of-speech tagging for each sentence
    tagged_sentences = [pos_tag(sentence) for sentence
in tokenized_sentences]

    named_entities = []

    # Named Entity Recognition using ne_chunk
    for tagged_sentence in tagged_sentences:
        tree = ne_chunk(tagged_sentence, binary=True)
        for subtree in tree:
            if isinstance(subtree, nltk.Tree) and
subtree.label() == 'NE':
                # Joining the words in the entity
                entity = ' '.join([word for word, _ in
subtree])
                named_entities.append(entity)

    return named_entities

def machine_learning_sentiment(text):
    # Sentiment Analysis using SentimentIntensityAnalyzer
    sia = SentimentIntensityAnalyzer()
    sentiment_scores = sia.polarity_scores(text)

    if sentiment_scores['compound'] > 0.05:
        return 'Positive'
    elif sentiment_scores['compound'] < -0.05:
        return 'Negative'
    else:
        return 'Neutral'

def hybrid_nlp_system(text):
    # Rule-based Named Entity Recognition
    named_entities = rule_based_ner(text)
    print("Named Entities:", named_entities)

    # Machine Learning Sentiment Analysis
    sentiment = machine_learning_sentiment(text)
    print("Sentiment:", sentiment)

# Example test text
example_text = "Apple Inc. is planning to launch a new
product next month. The company's stocks have been
rising recently."
```

```
# Apply the hybrid NLP system to the example text
hybrid_nlp_system(example_text)

% Output:

Named Entities: ['Apple Inc.']
Sentiment: Neutral
```

5.4 Introduction to Deep Learning in NLP

NLP has undergone a revolution thanks to deep learning, a kind of machine learning that makes it possible to generate and understand language in more complex and efficient ways. The way NLP tasks are tackled has changed dramatically as a result of deep learning. Significant progress has been made in language production and interpretation, as well as a variety of other NLP applications, thanks to the capacity to automatically extract intricate trends and representations from data.

Handcrafted rules and superficial linguistic elements were frequently used in traditional NLP techniques. However, deep learning enables models to automatically extract complex patterns and semantic correlations from raw data to create hierarchical representations of language.

Neural Networks in NLP:

Neural network topologies are used in deep learning to model intricate relationships in data. RNNs, transformer topologies, and Long Short-Term Memory networks (LSTMs) are frequently utilized in NLP.

Word Embeddings:

Word embeddings are dense vector representations of words in continuous vector spaces that were first introduced by deep learning. Semantic links between words have been captured in large part by methods like Word2Vec and GloVe.

Sequence Modeling:

Sequence modeling is a strong suit for deep learning models, especially those with recurrent and attention-based architectures. They are useful for tasks like text synthesis, summarization, and machine translation because they can comprehend and produce word sequences.

Transformers and Attention Mechanism:

A key component of NLP is the transformer architecture, which was first presented by Vaswani et al. It makes use of self-attention processes, which enable the model to assess the relative relevance of several phrases in a sequence and successfully capture long-range dependencies.

Pre-trained Models and Transfer Learning:

Pre-trained language models like Bidirectional Encoder Representations from Transformers (BERT) and Generative Pre_ trained Transformer (GPT) have become popular in deep learning. Transfer learning is made possible by these models' ability to be adjusted for particular NLP tasks after being trained on enormous datasets.

Natural Language Understanding:

Natural language comprehension challenges have greatly improved because to deep learning. It makes it possible for machines to under-stand the subtleties of language through semantic role labeling, named entity identification, and sentiment analysis.

Contextual Representations:

In contrast to conventional techniques, contextual information can be captured by deep learning models. As demonstrated by models such as Embeddings from Language Models (ELMo) contextual word embeddings offer representations that change according to the word's context inside a phrase.

Applications in Dialogue Systems and Chatbots:

Chatbots and intelligent dialogue systems have been made possible thanks in large part to deep learning. Models are able to comprehend user intent, produce responses that are pertinent to the situation, and have more organic interactions.

Challenges and Advances:

Considering the achievements, there are still issues with interpret-ability, explainability, and the requirement for a lot of labeled data. In order to overcome these obstacles and improve the potential of deep learning in NLP, research is still being conducted.

Implementation

```
% Python program

import tensorflow as tf
import numpy as np
```

```
from tensorflow.keras.preprocessing.text import
Tokenizer
from tensorflow.keras.preprocessing.sequence import
pad_sequences
from tensorflow.keras.models import Sequential
from tensorflow.keras.layers import Embedding, LSTM,
Dense

# Example sentences and Labels for sentiment analysis
sentences = ["I love this product!", "The movie was
disappointing.", "The weather is nice today."]
labels = np.array([1, 0, 1])  # Labels: 1 for
positive, 0 for negative sentiment

# Tokenize the sentences
tokenizer = Tokenizer(oov_token="<OOV>")  # oov_token
is used for out-of-vocabulary words
tokenizer.fit_on_texts(sentences)  # Build the word
index based on the sentences
word_index = tokenizer.word_index  # Get the word
index dictionary

# Convert sentences to sequences (tokenizing the
sentences)
sequences = tokenizer.texts_to_sequences(sentences)

# Pad sequences to ensure they have the same length
(pad shorter sequences and truncate longer ones)
padded_sequences = pad_sequences(sequences, maxlen=10,
padding='post', truncating='post')

# Build a simple LSTM model for sentiment analysis
model = Sequential([
    Embedding(len(word_index) + 1, 16, input_
length=10),  # Embedding layer
    LSTM(64),  # LSTM layer with 64 units
    Dense(1, activation='sigmoid')  # Output layer
with sigmoid activation (binary classification)
])

# Compile the model
model.compile(optimizer='adam', loss='binary_
crossentropy', metrics=['accuracy'])
```

```
# Train the model

model.fit(padded_sequences, labels, epochs=10)
# Make predictions on new data
new_sentences = ["I feel great about this!", "It's a
terrible experience."]
new_sequences = tokenizer.texts_to_sequences(new_
sentences)  # Convert new sentences to sequences
new_padded_sequences = pad_sequences(new_sequences,
maxlen=10, padding='post', truncating='post')  # Pad
the sequences

# Get predictions
predictions = model.predict(new_padded_sequences)

# Print the predictions
for i, sentence in enumerate(new_sentences):
    sentiment = "Positive" if predictions[i] > 0.5
else "Negative"
    print(f"Sentence: {sentence}\nSentiment:
{sentiment}\n")

% Output:
```

```
    Epoch 1/10
    1/1_____ 3s 3s/step - accuracy: 0.6667 - loss: 0.6925
    Epoch 2/10
    1/1_____ 0s 57ms/step - accuracy: 0.6667 - loss: 0.6896
    Epoch 3/10
    1/1_____ 0s 51ms/step - accuracy: 0.6667 - loss: 0.6867
    Epoch 4/10
    1/1_____ 0s 139ms/step - accuracy: 0.6667 - loss: 0.6838
    Epoch 5/10
    1/1_____ 0s 59ms/step - accuracy: 0.6667 - loss: 0.6808
    Epoch 6/10
    1/1_____ 0s 50ms/step - accuracy: 0.6667 - loss: 0.6776
    Epoch 7/10
    1/1_____ 0s 52ms/step - accuracy: 0.6667 - loss: 0.6743
    Epoch 8/10
    1/1_____ 0s 58ms/step - accuracy: 0.6667 - loss: 0.6707
    Epoch 9/10
    1/1_____ 0s 59ms/step - accuracy: 0.6667 - loss: 0.6669
    Epoch 10/10
    1/1_____ 0s 61ms/step - accuracy: 0.6667 - loss: 0.6629
    1/1_____ 0s 201ms/step
    Sentence: I feel great about this!
    Sentiment: Positive

    Sentence: It's a terrible experience.
    Sentiment: Positive
```

5.5 Applications of NLP

There are numerous uses for NLP in a variety of fields. Here are some common and notable applications:

- **Sentiment Analysis:** Textual analysis of attitudes and opinions, frequently utilized for social media monitoring, product evaluations, and customer feedback.
- **Text Summarization:** Creating succinct and logical summaries of longer texts is helpful for rapidly comprehending the key ideas in papers, articles, or conversations.
- **Language Translation:** Translating texts between languages, promoting interlanguage dialogue, and removing linguistic obstacles.
- **Speech Recognition:** Transcribing audible words into written form for usage in voice assistants, voice-activated gadgets, and transcription services.
- **Chatbots and Virtual Assistants:** Creating interactive conversational agents that can understand and respond to user queries, providing customer support or information retrieval.
- **Text Classification:** Categorizing documents or text into predefined classes, such as spam detection, topic classification, and sentiment categorization.
- **Text Generation** [10]: Generating human-like text, which can be used for content creation, creative writing, and even in the development of automated storytelling.
- **Spell and Grammar Checking** [11]: Correcting spelling and grammar mistakes in text, improving the quality and readability of written content.

5.5.1 Sentiment Analysis

One important use of NLP is sentiment analysis, which analyzes and ascertains the emotional tone conveyed in a text by means of algorithms. It seeks to determine if the text's sentiment is neutral, negative, or positive. This technology is frequently used to comprehend social media material, customer feedback, and public opinion. It gives organizations, marketers, and decision-makers important insights to assess the general sentiment and react appropriately.

Implementation

Algorithm

1. **Import Libraries:** Import the necessary libraries, such as nltk and the SentimentIntensityAnalyzer class.
2. **Initialize Sentiment Intensity Analyzer:** Create an instance of the SentimentIntensityAnalyzer.
3. **Analyze Sentiment:** To obtain sentiment ratings (positive, negative, neutral, and complex) for a given text, use the polarity_scores method.
4. **Interpret Results:** Interpret the compound score to classify the sentiment.

```
% Python program

import nltk
from nltk.corpus import stopwords
from nltk.tokenize import word_tokenize
from nltk.sentiment import SentimentIntensityAnalyzer

# Preprocess the text
def preprocess_text(text):
    # Tokenize the text
    words = word_tokenize(text)

    # Remove stopwords
    stop_words = set(stopwords.words('english'))

    # Filter words: keep only alphanumeric words that
are not stopwords
    filtered_words = [word.lower() for word in words
if word.isalnum() and word.lower() not in stop_words]

    # Join the filtered words back into a string
    preprocessed_text = ' '.join(filtered_words)

    return preprocessed_text

# Analyze sentiment of the text
def analyze_sentiment(text):
    # Preprocess the text
    preprocessed_text = preprocess_text(text)
```

```
# Initialize sentiment analyzer
sia = SentimentIntensityAnalyzer()

# Get sentiment scores
sentiment_scores = sia.polarity_scores
(preprocessed_text)

    # Classify sentiment based on the compound score
    if sentiment_scores['compound'] >= 0.05:
        sentiment = 'Positive'
    elif sentiment_scores['compound'] <= -0.05:
        sentiment = 'Negative'
    else:
        sentiment = 'Neutral'

    return sentiment

# Example usage
product_review = "I absolutely love this product! It's
amazing."
sentiment_result = analyze_sentiment(product_review)
print(product_review)
print(f"Sentiment: {sentiment_result}")

% Output:

I absolutely love this product! It's amazing.
Sentiment: Positive
```

5.5.2 *Prediction of Next Word*

The prediction of the next word is an application of NLP that involves using algorithms to predict the most likely word that would follow a given sequence of words in a sentence or text [12–15]. This application typically relies on language models, such as recurrent neural networks (RNNs) or transformer models like GPT. By training on large datasets, these models learn the patterns and relationships between words, allowing them to make intelligent predictions about the next word based on context. This technology is commonly used in autocomplete suggestions, text completion, and predictive typing applications to enhance user experience and assist in generating coherent and contextually appropriate text.

Implementation

```
% Python program

class NextWordPredictor:

    def __init__(self, corpus):
        # Initialize with the given corpus
        self.corpus = corpus
        # Initialize the word frequency dictionary
        self.word_frequency = {}
        # Build the word frequency table
        self.build_word_frequency()

    def build_word_frequency(self):
        # Build word frequency based on the corpus
        for sentence in self.corpus:
            words = sentence.split()
            for i in range(len(words) - 1):  # Loop
through words except the last one
                current_word = words[i]
                next_word = words[i + 1]
                if current_word not in self.
word_frequency:
                    self.word_frequency[current_word]
= {}
                if next_word not in self.word_frequency
[current_word]:
                    self.word_frequency[current_word]
[next_word] = 1
                else:
                    self.word_frequency[current_word]
[next_word] += 1

    def predict_next_word(self, current_word):
        # Predict the next word given the current word
        if current_word in self.word_frequency:
            next_words = self.
word_frequency[current_word]
            # Predict the next word by finding the
most frequent one
            predicted_word = max(next_words, key=next_
words.get)
```

```
            return predicted_word
        else:
            return None


# Example Usage
corpus = [
    "Natural language processing is a subfield of
artificial intelligence.",
    "It focuses on the interaction between computers
and humans using natural language.",
    "NLP applications include machine translation,
sentiment analysis, and speech recognition."
]

predictor = NextWordPredictor(corpus)

input_word = "language"
predicted_next_word = predictor.predict_next_word
(input_word)

if predicted_next_word:
    print(f"The predicted next word after '{input_
word}' is '{predicted_next_word}'.")
else:
    print(f"No prediction available for the word
'{input_word}'.")

% Output:

The predicted next word after 'language' is
'processing'.
```

5.6 Summary

This chapter explores the role of AI in NLP, highlighting how machine learning and deep learning techniques enhance language understanding and processing. It begins by discussing the integration of AI in NLP, followed by an overview of Supervised and Unsupervised Machine Learning techniques, which enable automated text classification, sentiment analysis, and language modeling. Hybrid ML systems, which combine multiple learning approaches for improved NLP performance, are also introduced. This chapter further examines deep learning for NLP, covering its applications in handling complex linguistic tasks such as next-word prediction, sentiment analysis, and information extraction. It also explores real-world case studies, including Language Understanding Intelligent Service (LUIS) for natural language comprehension. Additionally, it discusses advanced NLP applications such as text summarization, text-to-speech conversion, and chatbots, which have revolutionized human-computer interactions. Overall, this chapter provides insights into how AI-driven approaches are shaping the future of NLP, enabling machines to process, analyze, and generate human language with greater accuracy and efficiency.

References

1. M. Wankhade, A. C. S. Rao, and C. Kulkarni, "A survey on sentiment analysis methods, applications, and challenges," *Artif. Intell. Rev.*, vol. 55, no. 7, pp. 5731–5780, 2022.
2. J. Li, A. Sun, J. Han, and C. Li, "A survey on deep learning for named entity recognition," *IEEE Trans. Knowl. Data Eng.*, vol. 34, no. 1, pp. 50–70, 2020.
3. K. Kowsari, K. Jafari Meimandi, M. Heidarysafa, S. Mendu, L. Barnes, and D. Brown, "Text classification algorithms: A survey," *Information*, vol. 10, no. 4, p. 150, 2019.
4. H. Zhao, D. Phung, V. Huynh, Y. Jin, L. Du, and W. Buntine, "Topic modelling meets deep neural networks: A survey," *arXiv Prepr. arXiv2103.00498*, 2021.
5. A. Petukhova, J. P. Matos-Carvalho, and N. Fachada, "Text clustering with large language model embeddings," *arXiv Prepr. arXiv2403.15112*, 2024.
6. S. J. Johnson, M. R. Murty, and I. Navakanth, "A detailed review on word embedding techniques with emphasis on word2vec," *Multimed. Tools Appl.*, vol. 83, no. 13, pp. 37979–38007, 2024.
7. W. S. El-Kassas, C. R. Salama, A. A. Rafea, and H. K. Mohamed, "Automatic text summarization: A comprehensive survey," *Expert Syst. Appl.*, vol. 165, p. 113679, 2021.
8. A. M. N. Allam and M. H. Haggag, "The question answering systems: A survey," *Int. J. Res. Rev. Inf. Sci.*, vol. 2, no. 3, p. 66, 2012.
9. A. Rajput, "Natural language processing, sentiment analysis, and clinical analytics," in M. D. Lytras and A. Sarirete (eds.) Innovation in Health Informatics, Elsevier, Amsterdam, Netherlands, 2020, pp. 79–97.
10. T. Iqbal and S. Qureshi, "The survey: Text generation models in deep learning," *J. King Saud Univ. Inf. Sci.*, vol. 34, no. 6, pp. 2515–2528, 2022.

11. A. Fahda and A. Purwarianti, "A statistical and rule-based spelling and grammar checker for Indonesian text," in *2017 International Conference on Data and Software Engineering (ICODSE)*, Palembang, Indonesia, 2017, pp. 1–6.

12. J. Stremmel and A. Singh, "Pretraining federated text models for next word prediction," in *Advances in Information and Communication: Proceedings of the 2021 Future of Information and Communication Conference (FICC)*, Vancouver, Canada, *Volume 2*, 2021, pp. 477–488.

13. R. Sharma, N. Goel, N. Aggarwal, P. Kaur, and C. Prakash, "Next word prediction in hindi using deep learning techniques," in *2019 International conference on data science and engineering (ICDSE)*, Patana, India, 2019, pp. 55–60.

14. M. Soam and S. Thakur, "Next word prediction using deep learning: A comparative study," in *2022 12th International Conference on Cloud Computing, Data Science & Engineering (Confluence)*, Noida, India, 2022, pp. 653–658.

15. A. Rianti, S. Widodo, A. D. Ayuningtyas, and F. B. Hermawan, "Next word prediction using lstm," *J. Inf. Technol. Its Util.*, vol. 5, no. 1, p. 432033, 2022.

Index

For Product Safety Concerns and Information please contact our EU
representative GPSR@taylorandfrancis.com
Taylor & Francis Verlag GmbH, Kaufingerstraße 24, 80331 München, Germany